Complexity and Creative Capacity

Complexity theories gained prominence in the 1990s with a focus on self-organising and complex adaptive systems. Since then, complexity theory has become one of the fastest-growing topics in both the natural and social sciences, and touted as a revolutionary way of understanding the behaviour of complex systems.

This book uses complexity theory to surface and challenge the deeply held cultural assumptions that shape how we think about reality and knowledge. In doing so it shows how our traditional approaches to generating and applying knowledge may be paradoxically exacerbating some of the 'wicked' environmental problems we are currently facing. The author proposes an innovative and compelling argument for rejecting old constructs of knowledge transfer, adaptive management and adaptive capacity. The book also presents a distinctively coherent and comprehensive synthesis of cognition, learning, knowledge and organising from a complexity perspective. It concludes with a reconceptualisation of the problem of knowledge transfer from a complexity perspective, proposing the concept of creative capacity as an alternative to adaptive capacity as a measure of resilience in socio-ecological systems.

Although written from an environmental management perspective, it is relevant to the broader natural sciences and to a range of other disciplines, including knowledge management, organisational learning, organisational management, and the philosophy of science.

Kelly Chapman is a consultant environmental scientist, based in Calgary, Canada. She has a PhD from Edith Cowan University in Western Australia and more than twenty years of practical experience in environmental management and planning, environmental communications, and biophysical projects in Africa, Australia and Canada.

Routledge Explorations in Environmental Studies

Nuclear Waste Management and Legitimacy
Nihilism and responsibility
Mats Andrén

Nuclear Power, Economic Development Discourse and the Environment
The case of India
Manu V. Mathai

Federalism of Wetlands
Ryan W. Taylor

Governing Sustainable Urban Renewal
Partnerships in action
Rory Shand

Employee Engagement with Sustainable Business
How to change the world whilst keeping your day job
Nadine Exter

The Political Economy of Global Warming
The terminal crisis
Del Weston

Communicating Environmental Patriotism
A rhetorical history of the American environmental movement
Anne Marie Todd

Environmental Justice in Developing Countries
Perspectives from Africa and Asia-Pacific
Rhuks Temitope Ako

Climate Change and Cultural Heritage
A race against time
Peter F. Smith

A Global Environmental Right
Stephen Turner

The Coming of Age of the Green Community
My neighbourhood, my planet
Erik Bichard

Fairness and Justice in Environmental Decision Making
Water under the bridge
Catherine Gross

Carbon Politics and the Failure of the Kyoto Protocol
Gerald Kutney

Trade, Health and the Environment
The European Union put to the test
Marjolein B.A. van Asselt, Michelle Everson and Ellen Vos

Conflict, Negotiations and Natural Resource Management
A legal pluralism perspective from India
Edited by Maarten Bavinck and Amalendu Jyotishi

Philosophy of Nature
Rethinking naturalness
Svein Anders Noer Lie

Urban Environmental Stewardship and Civic Engagement
How planting trees strengthens the roots of democracy
Dana R. Fisher, Erika S. Svendsen and James J.T. Connolly

Disaster Risk Reduction for Economic Growth and Livelihood
Investing in resilience and development
Edited by Ian Davis, Kae Yanagisawa and Kristalina Georgieva

Energy Security Cooperation in Northeast Asia
Edited by Bo Kong and Jae Ku

Water Politics and Spiritual Ecology
Custom, environmental governance and development
Lisa Palmer

The Politics of Ecosocialism
Transforming welfare
Edited by Kajsa Borgnäs, Teppo Eskelinen, Johanna Perkiö and Rikard Warlenius

Peak Energy Demand and Demand Side Response
Jacopo Torriti

The Green Economy in the Gulf
Edited by Mohamed Abdelraouf and Mari Luomi

Ecology, Sustainable Development and Accounting
Seleshi Sisaye

Environmental Change and the World's Futures
Ecologies, ontologies and mythologies
Edited by Jonathan Paul Marshall and Linda H. Connor

Environmental Adaptation and Eco-cultural Habitats
A coevolutionary approach to society and nature
Johannes Schubert

Complexity and Creative Capacity
Rethinking knowledge transfer, adaptive management and wicked environmental problems
Kelly Chapman

Contemporary Perspectives on Ecofeminism
Mary Phillips and Nick Rumens

Conservation and Culture
Beyond anthropocentrism
Eleanor Shoreman-Ouimet and Helen Kopnina

Complexity and Creative Capacity

Rethinking knowledge transfer, adaptive management and wicked environmental problems

Kelly Chapman

Routledge
Taylor & Francis Group
LONDON AND NEW YORK

earthscan
from Routledge

First published 2016
by Routledge

2 Park Square, Milton Park, Abingdon, Oxfordshire OX14 4RN
711 Third Avenue, New York, NY 10017

Routledge is an imprint of the Taylor & Francis Group, an informa business

First issued in paperback 2017

British Library Cataloguing-in-Publication Data
A catalogue record for this book is available from the British Library

Library of Congress Cataloging in Publication Data
Names: Chapman, Kelly, 1967-
Title: Complexity and creative capacity : rethinking knowledge transfer, adaptive management and wicked environmental problems / Kelly Chapman.
Description: Abingdon, Oxon ; New York, NY : Routledge is an imprint of the Taylor & Francis Group, an Informa business, [2015] | Series: Routledge explorations in environmental studies | Includes bibliographical references.
Identifiers: LCCN 2015021214
Subjects: LCSH: Environmental management. | Knowledge management. | Adaptive natural resource management. | Complexity (Philosophy)
Classification: LCC GE300 .C43 2015 | DDC 333.701—dc23
LC record available at http://lccn.loc.gov/2015021214

ISBN: 978-1-138-92999-9 (hbk)
ISBN: 978-0-8153-9539-3 (pbk)

Typeset in Times New Roman
by FiSH Books Ltd, Enfield.

Dedicated to the memory of
Ken and Betty Chapman
and Chris Keyes

Dedicated to the memory of
Ken and Betty Chapman
and Doris Keyes

Contents

Acknowledgments

This book would not have been possible without the support of many people who contributed ideas, support, and encouragement. First and foremost, many thanks to the people who took time out of their busy schedules to meet with me and provide their valuable input for the study that inspired this book, who are too many to list. Thank you to Commonwealth Scientific and Industrial Research Organisation (CSIRO) and Edith Cowan University for their financial contributions to the study, and to the Western Australia Department of Parks and Wildlife, the Ningaloo Turtle Program, and the Gascoyne Development Commission for providing in-kind support. Special thanks to all those who opened up their homes and offered me places to stay while I conducted my research, notably Milton Landon, Lyn Irvine, Susie Bedford, Paul Richardson, Peter Dellar, Kylie West, Colleen Sims, Karen Hattingh and Mike Ross. I also extend my appreciation to the many people who gave me encouragement, moral support and friendship along the way, among them my dissertation supervisors Pierre Horwitz, Pascal Scherrer and Geoff Syme; my study collaborators Tod Jones, Beth Fulton, Karen Thompson and Fabio Boschetti; Scott Gardner and Annabel Mazzella, for your ongoing support and ideas; my partner in kiting, Tania Knight; my wonderful climbing friends in Perth; my 're-discovered' family and friends at the Lake; and my long-neglected Canadian friends and family—thanks for sticking with me through this process. Thanks also to Tina Holgate and Teresa Chapman for the onerous task of reviewing and editing the document. And finally eternal gratitude to Norma Foran, Bruce Chapman, Teresa Chapman, and Maggie Wheeler—I couldn't have done it without your loving support; and to Chris and Patricia for illuminating the path.

1 Introduction

The relationship between knowledge and adaptive capacity

> The problems we have created in the world today will not be solved by the level of thinking that created them.
>
> (Albert Einstein[1])

Introduction

A few years ago, I found myself puzzling over the disappointing results of a recent undertaking: a knowledge transfer project aimed at encouraging local stakeholders to apply research and modelling results from a multi-million-dollar research project in an environmentally sensitive landscape. We had spent hundreds of hours hosting dozens of workshops, presentations and meetings around the region. But despite this considerable investment, a post-project evaluation showed our efforts had only limited effect on local knowledge levels and research application.

This led me to take stock on my 20-year career as an environmental scientist: three continents' worth of experience in environmental planning, management, knowledge transfer and communication. What struck me on reflection was how few of these endeavours actually resulted in substantial on-the-ground impact. As I began to investigate the literature for possible reasons, what struck me further was the ubiquity of my experience. Finding ways of continuously generating, sharing and applying knowledge from a diversity of disciplines and viewpoints is a cornerstone of adaptive management and building the adaptive capacity (i.e. resilience) of governance systems such that they are able to detect and respond to environmental change (Folke *et al.* 2005; Holling 1978). Failing to do so leaves us unprepared for ecological crises (Gunderson 2003)—such as collapses in commercial fisheries, catastrophic flooding and fires, and devastating pest invasions—and vulnerable to their often irreversible and long-reaching ecological and socio-economic impacts. Indeed, establishing stronger feedback loops between decision-makers and ecological systems is now widely regarded as one of the greatest challenges to global sustainability (Millennium Ecosystem Assessment 2005). Yet disturbingly, it appears that two of the standard mechanisms for doing so—knowledge transfer and adaptive management—are falling short at alarming rates (Medema *et al.* 2008; Roux *et al.* 2006).

Resilience and adaptive management

It is now widely accepted that ecological and social systems are complex, interconnected, non-linear and unpredictable entities that cannot be understood from a single perspective or discipline (be it ecological, economic or social) or abstracted from their historical, social, political or landscape contexts (e.g. Funtowicz and Ravetz 1993; Meppem and Bourke 1999). The complexity and unpredictability of ecological and social systems clearly present substantial challenges for today's resource managers, who must make difficult decisions based on incomplete information in an uncertain world. In such environments, resource managers are faced with so-called *wicked problems*—complex socio-environmental problems that span multiple systems, disciplines and worldviews. According to Rittel and Webber (1973) wicked problems are impossible or difficult to solve because they cannot be singularly defined and do not have right or wrong solutions (just better or worse as subjectively defined by involved stakeholders). In addition, wicked problems have numerous subjective causes and their implemented 'solutions' have significant consequences, meaning there is no opportunity for trial and error learning. As Brown *et al.* (2010) note, such problems are not morally wicked; rather they are diabolical in their resistance to being solved.

Conventional scientific/technical approaches to resource management based on linear causality and reductionism have neglected the agenda of social change which parallels the process of defining environmental problems, instead emphasising the more easily dealt with technical and scientific aspects. Meppem and Bourke (1999) argue that this reliance on 'technofix' rather than social approaches to environmental problems has resulted in impotent politics, ineffective environmental policy, and an inoperable concept of sustainability. Centralised institutions using models based on linear thinking and mechanistic views of nature to increase resource production and efficiency have eroded the variation and resilience of natural systems, making them vulnerable to crisis (Carpenter *et al.* 2002; Holling and Meffe 1996). Many are now arguing that conventional science and management are ill equipped to deal with the 'wickedness' of complex socio-environmental problems, and that new *resilience*-based approaches are needed, such as *adaptive management.*

Resilience, adaptive capacity and adaptive management are concepts that were introduced to environmental management by Buzz Holling in 1973, and since built on by researchers with the Resilience Alliance programme. In his work, Holling blended systems theory with ecology to introduce the notion of resilience and cycles of destruction and renewal in ecological systems (Holling 1973). He also introduced the concept of adaptive management (Holling 1978). Holling's ideas were later expanded into *panarchy theory*, which looks at resilience and adaptive cycles in socio-ecological systems (Gunderson and Holling 2002). Walker *et al.* (2002: 6) define resilience as "the potential of a system to remain in a particular configuration and to maintain its feedbacks and functions, and involves the ability of the system to reorganise following disturbance-driven change". As such, 'resilience', as applied to ecosystems and socio-ecological systems, has three defining characteristics:

- the amount of change the system absorbs while still retaining the same controls on function and structure;
- the degree to which the system can self-organise (i.e. organise itself without centralised or outside direction or control); and
- the ability to increase the capacity for learning and adaptation.

(Berkes *et al.* 2003)

According to the Resilience Alliance (2015), systems with high adaptive capacity are able to respond to change by re-configuring themselves without significant losses in critical functions, such as primary productivity, hydrology, or socio-economic health. The adaptive capacity of *ecological* systems is linked to genetic, biological and landscape diversity (Bengtsson *et al.* 2003; Carpenter *et al.* 2001; Peterson *et al.* 1998). The adaptive capacity of *social* systems, on the other hand, is linked to the presence of flexible and long-enduring institutions that respond to challenges by experimenting, learning, problem solving, and balancing the interests of various stakeholder groups (Scheffer *et al.* 2000; Walker *et al.* 2002). Panarchy theory seeks to understand the complex cross-scale dynamics between social and ecological processes, and how they build or erode socio-ecological adaptive capacity over time and space (Gunderson and Holling 2002). A consequence of losing resilience, and therefore adaptive capacity, is the reduced ability to deal with change because of there being fewer options for renewal and reorganisation after periods of disturbance or crisis (Resilience Alliance 2015).

Adaptive management is a framework through which research, policy and local practice can be integrated to increase the adaptive capacity of systems through cyclical learning and policy adjustment and implementation (Holling 1978; Walters 1986). This approach allows testing of key hypotheses about the system by comparing the outcomes of different policies and practices. Berkes *et al.* (2003) note that adaptive management is founded on social and institutional learning using feedback from the environment to shape policy, followed by experimentation to further shape subsequent policy. Because socio-ecological systems are self-organising systems that are always in flux, our understanding of them needs to be continuously updated and adjusted (Carpenter and Gunderson 2001). In sum, sustainable governance of socio-ecological systems requires flexible institutions that maintain their responsiveness to environmental feedback through continuous learning and experimentation (Folke *et al.* 2005).

The reality, however, is that adaptive management is very resource and time intensive, especially when it involves engaging multiple stakeholder groups (Huitema *et al.* 2009). Indeed, its application may require complete overhaul of existing management and decision-making systems (Berkes 2010). Consequently, it appears that few such projects exist in real life, and those which do are often unsuccessful in terms of improving sustainable resource management (Curtin 2014; Plummer *et al.* 2012). Reviews show that around 70 per cent of adaptive and integrated resource management projects fail on implementation (e.g. Medema *et al.* 2008), with Walters (1997) citing that a staggering 23 out of 25 major adaptive management planning exercises he'd been involved with had 'vanished' without visible product or became trapped in an endless cycle of model refinement. Medema *et al.* (2008) conclude that the barriers to institutional reform are considerable, and should not be underestimated.

Knowledge transfer

Knowledge and learning play a pivotal role in adaptive management. The wicked problems presented by the global biodiversity and climate change crises have heightened the imperative for knowledge integration and sharing between a diversity of disciplines and stakeholders (Cash *et al.* 2003; Folke *et al.* 2003; Raymond *et al.* 2010). This is exemplified by increasing calls for multi-stakeholder resource management approaches based on collaborative learning, such as communicative planning (e.g. Meppem and Bourke 1999; Wondolleck and Yaffee 2000) and adaptive co-management (e.g. Armitage *et al.* 2008; Olsson *et al.* 2004). Roux *et al.* (2006) state that sustainable ecosystem management relies on a broad base of knowledge that is continuously updated to reflect current issues and needs, and which flows from scientists to managers with minimal delay. However, many now argue that the main barrier to sustainability is no longer lack of knowledge; it's the failure to *act* on that knowledge (e.g. Arlettaz *et al.* 2010; Ehrlich and Kennedy 2005; Fischer *et al.* 2012).

Knowledge transfer refers to the spectrum strategies used to bridge the gap between research, policy and management (Roux *et al.* 2006). Examples of such strategies include: generating awareness and understanding by disseminating information, building support and involvement through workshops and meetings, and securing commitment through group problem solving. Knowledge transfer is optimised when research informs practice, and practice in turn informs research—it is not successful unless it results in application of research in practice (i.e. *action*; *ibid.*). This is increasingly recognised as requiring strong linkages and interpersonal interaction between scientists and managers/decision-makers (Chapman *et al.* submitted). Yet despite this recognition and mounting imperative, knowledge transfer efforts often still fall short of seeing research applied on the ground (Roux *et al.* 2006). Roux *et al.* (*ibid.*) attribute this to a legacy of disciplinary fragmentation that has divorced scientific knowledge from its application, to the detriment of natural resource management. They also relate it to cultural clashes between resource managers and scientists. McNie (2007) suggests that the problem is exacerbated by a lack of understanding about knowledge and decision systems, and how to improve on decision processes associated with environmental decision-making. The US National Research Council (1999) has partially attributed this lack of understanding to the failure of those in the natural sciences to appreciate the contributions of the social and behavioural sciences. This seems to be a considerable oversight, given that most of the barriers to improving the adaptive capacity of institutions are related to *social* processes operating both within and outside of these human systems.

Social science perspective

Environmental scientists have traditionally held a relatively simplistic view of knowledge transfer as a mechanical process of transferring research findings from scientists to managers and management institutions. The focus of such knowledge transfer efforts has largely been on improving:

- salience, credibility and legitimacy of research and modelling results;
- interpretation of research/modelling results into information that is relevant and compelling to target audiences/users;
- distribution of generated information/models to the 'right' audiences/users;
- institutionalisation of information/models in organisational memory and structures (i.e. policy and procedures).

This approach is based on what Huber (1991) describes as a behaviourist approach to learning. It is premised on the assumption that if an entity receives and processes information, its behaviour will change. However, the simplicity of this assumption has long been disputed in the social sciences. For decades, issues around cognition, organisational learning and the effective sharing and application of knowledge have occupied social science scholars (e.g. Argyris and Schön 1978; Festinger 1957; Nonaka and Konno 1998; Senge 1990; Snowden 2003; Wenger 1998). Learning and behaviour change are the result of complex cognitive and social processes. Rarely does the presentation of new information on its own, regardless of how compelling, cause substantial changes in individual or organisational behaviours (Argyris and Schön 1974; Markus and Zajonc 1985; McKenzie-Mohr 2011). Indeed, research on cognitive dissonance has shown that presenting people with new information can actually cause them to entrench their existing behaviours and more strongly resist further change (Nyhan and Reifler 2010). Likewise, learning that does not result in major shifts in a learner's beliefs and underlying assumptions typically fails to change their behaviour in substantial ways, indeed often *escalates* errors being committed (Watzlawick *et al.* 2011).

In addition, knowledge management scholars draw an important distinction between information and knowledge, a distinction largely unrecognised by those in the environmental sciences. Knowledge contains an unspoken *tacit* component that is linked to people's relationships and capacity to act (Nonaka and Takeuchi 1995), whereas information, or *explicit* knowledge, seldom results in behaviour change on its own. The social sciences also present a large body of work describing preconditions for learning and behaviour change, which is largely unrecognised by those in the environmental sciences. For example, learning and behaviour change can be enhanced by free and open dialogue, trust and relationships, sharing control and responsibility, minimising defensiveness, crisis and destabilisation and deep reflection (see Argyris and Schön 1996; Bartunek and Moch 1987; Hedberg *et al.* 1976; Senge 1990; Wenger 1998). Deploying a knowledge broker or change agent can also help catalyse the social interactions needed for organisational learning to occur (e.g. Bartunek and Moch 1987; Michaels 2009).

Knowledge transfer is also challenging because of the complex nature of human social systems. Complexities and difficulties multiply when attempting to transfer knowledge between organisations because of the "multifaceted nature of the boundaries, cultures, and processes involved" (Easterby-Smith *et al.* 2008: 677). This makes it impossible to devise a standard recipe for engaging stakeholders in a knowledge transfer process. The composition, influence, knowledge, motivations and actions of stakeholders differ for any given place and for any given time, with groups and people forming complex and ever-changing webs of relationships that are inherently unpredictable (Fulton *et al.* 2013). This capacity of human systems to spontaneously change in unpredictable ways limits our ability to design and plan a knowledge transfer process. In addition, the researchers themselves affect the social system—as soon as they begin interacting with stakeholders, stakeholder perceptions, knowledge and actions begin to change in response (see Juarrero 1999; Stacey *et al.* 2000).

Clearly, the social sciences have much to contribute to our understanding of knowledge transfer in environmental management. The importance of using knowledge and feedback to improve adaptive capacity is not reserved for resource management agencies. To be competitive and successful, businesses must also be able to continuously learn and quickly transform their learning into action (i.e. they must be adaptive). This is the focus of the *organisational learning* and *knowledge management* fields (e.g. Argyris and Schön 1996; Nonaka and Takeuchi 1995). Easterby-Smith *et al.* (2008) state that 20 years of empirical research has shown that knowledge transfer within and between firms can improve their learning capacity and create competitive advantage. Many businesses aim to improve their learning and adaptive capabilities by undertaking strategic or change management initiatives. Senge (1990) refers to such businesses as *learning organisations*. Change management in the business world is widely premised on understandings and theories of organisational behaviour derived from the fields of organisational learning, knowledge management and complexity theory.

However, as is the case with adaptive management in the environmental world, it appears that most change management initiatives in the world of business and organisations likewise fail. Senge *et al.* (1999) cite failure rates of around 70 per cent, while Burnes (2005) cites rates of 80 per cent and higher. He remarks that despite decades of debate in the management literature around how to manage change, and despite mounting imperative to improve organisational adaptability in today's fast-moving business environment, there remains significant difficulty in successfully bringing it about. In their review of inter-organisational knowledge transfer Easterby-Smith *et al.* (2008) likewise conclude that knowledge transfer is a complex phenomenon that is rarely successful in practice, even *within* organisations.

Whereas the low success rates of knowledge transfer and adaptive management projects in the chronically under-resourced environmental world were relatively unsurprising to me, I was startled to see such a striking parallel in the business world. It occurred to me that the intractable difficulties associated with achieving successful knowledge transfer and adaptive management might run much deeper than a simple failure to integrate disciplinary learnings across silos. Indeed, the failure rates common to both the social and environmental sciences, despite their very different approaches, suggested a common root to the problem, one not yet adequately addressed in either arena. It seemed that adaptive management and knowledge transfer were not only frequently failing to resolve wicked problems, they were also wicked problems in and of themselves. I realised I needed to go back to the drawing board and re-evaluate these failures—including that of my own knowledge transfer project—in a completely different context.

Problem formulation and philosophy

The quote attributed to Albert Einstein at the head of this chapter is a reflection of Gödel's incompleteness theorem, whereby Gödel mathematically proved that:

> No system can prove its consistency within its own framework, proof can only come from outside, with additional axioms, premises, concepts, comparisons etc., which the system cannot prove, and which themselves are only provable by recourse to yet a wider framework.
>
> (Watzlawick *et al.* 2011: 24)

Gödel's theorem[2] established the essential limitations of axioms in arithmetic systems, by showing that within such systems there is always a statement, which although apparently true, cannot be proven within the system.[3] In other words, the theorem shows that there are problems which cannot be solved using the rules and premises that govern the system within which the problems are situated. Such problems can only be solved by adding new premises or 'rules' (i.e. new starting points for reasoning). Meppem and Bourke (1999) interpret Einstein's observation as indicating that the sustainability problems we have created are a result of certain dominant ways of thinking and knowing in Western society, which cannot therefore be used to successfully resolve such problems.

Conklin *et al.* (2007) argue that complexity is not necessarily the defining character of wicked problems; rather "it is about a fundamentally different kind of challenge … one that makes solution secondary and problem understanding central". In other words, the solution to the problem depends on how the problem is formulated (Rittel and Webber 1973). This relates to Watzlawick *et al.*'s (2011) seminal work *Change: Principles of Problem Formulation and Problem Resolution.* In their book the authors apply theorems from mathematical logic[4] (including Gödel's) to explain why problems arise and why they are perpetuated in some instances and resolved relatively easily in others. They outline how problems that appear to defy resolution are framed such that their assumptions preclude effective solution, and how these problems are unintentionally *reinforced* (and sometimes even created) by the very strategies applied to resolve them. As per Einstein's observation and Gödel's theorem, the solutions that can actually resolve such problems are *qualitatively* very different than those previously attempted because they necessarily stand outside the modes of thought that created the problem to begin with. Watzlawick *et al.* state that finding such solutions requires stepping back from the problem, examining our *assumptions* about the problem situation rather than the problem itself, then reframing the problem accordingly.

Thus, wicked problems call for a qualitatively different approach to conventional problem solving (Palmer *et al.* 2007), as no level of linear thinking can ever formulate their solution (Conklin *et al.* 2007). Watzlawick *et al.*'s principles of problem formulation suggest that in order to deal with wicked problems, we need to transcend the dominant modes of thought that lock us into patterns of behaviour and problem-solving that unintentionally perpetuate or escalate the problems we are trying to resolve. This involves standing back[5] and recognising some of the core assumptions that underpin Western society's understanding of reality and knowledge. These assumptions, which we unconsciously absorb and pass on, are so deeply rooted in the Western psyche that most people who are unfamiliar with metaphysics and epistemology (i.e. most people) are completely unaware that their perception of the world is largely shaped by these overarching assumptions (see Bateson 2002; Shotter 1994). Indeed, they are also unaware that knowledge and reality can be conceived in ways other than that dictated by Newtonian science-based paradigm that has prevailed in the West since the Victorian era. The famed psychologist and anthropologist Gregory Bateson writes:

> I have taught various branches of behavioural biology and cultural anthropology to American students ranging from college freshmen to psychiatric residents, in various schools and teaching hospitals, and I have encountered a very strange gap in their thinking that springs from a lack of certain tools of thought. This lack is rather equally distributed at all levels of education, among students of both sexes and among humanists as well as scientists. Specifically, it is a lack of knowledge about the presuppositions not only of science but of every day life.
>
> (Bateson 2002: 23)

A basic function of philosophy is to examine the assumptions that underlie our thinking (Heylighen *et al.* 2007). Lyotard (1984) states that whereas an expert knows what he does and does not know, and *concludes*, a philosopher knows neither, and instead *questions*. However, today's natural scientists—being experts—are largely unschooled in philosophy and its role in shaping history, science and our understanding of the world. This 'divorce' between science and philosophy is relatively recent, having occurred sometime after the Second World War.[6] Quantum physicist Carlo Rovelli (2012) remarks that it is fashionable today to discard philosophy, to believe that it is unnecessary because we have science. He argues that this is a naïve attitude, that our greatest scientists—Heisenberg, Einstein and Galileo, for example—made such momentous discoveries *because* they were fully versed in philosophy. Rovelli, among others (e.g. Bohm 2012; Noë and Thompson 2004; Peat 1990), is now making strong arguments that science, far from having freed itself from philosophy, is in dire need of its help.

The allergy scientists have acquired for philosophy is poignantly illuminated by John Maddox, senior editor for *Nature*, in his attack on Rupert Sheldrake's (1981) controversial theory of morphogenesis. In his theory, Sheldrake used the metaphysical thinking of philosopher and mathematician Alfred Whitehead to generate an alternative explanation for the holist nature of organisms. His work had received a mildly positive reception in *New Scientist* and among some quantum physicists before the furore raised by Maddox's attack (Freeman 2005). In a BBC interview Maddox (1994) stated that "Sheldrake's is not a scientific theory. Sheldrake is putting forward magic instead of science, and that can be condemned, in exactly the language that the Pope used to condemn Galileo, and for the same reasons: it is heresy".

Maddox's reference to Galileo drips with irony, as did the headline for his editorial in *Nature*: *A Book for Burning?* (Maddox 1981), wherein he states that Sheldrake's book "is the best candidate for burning there has been for many years" (*ibid.*: 245). The attack effectively ended Sheldrake's academic career as a respected plant physiologist (Freeman 2005). Indeed, according to Freeman (*ibid.*), the antagonism generated by Sheldrake's theory runs so deep that 20 years later journals still run the risk of ostracism if they publish Sheldrake's work.[7] Freeman attributes the scientific establishment's hysterical and dogmatic attacks on Sheldrake's work to a divisive moralism (the 'heretical' anti-science people vs. the 'good' pro-science people), and a commitment to the existing scientific paradigm that is so strong that explorations outside of it are rejected out of hand. Indeed the reaction to Sheldrake's theory is as predicted of 'normal' scientists in Kuhn's (1962) paradigmatic theory of scientific progress. According to Kuhn, normal scientists assume "that a paradigm provides the means for the solution of the puzzles posed within it", and are therefore uncritical of it (Chalmers 2002: 110). As such, observations or theories that fail to fit the dominant paradigm are often discredited or dismissed as anomalies by the normal scientific community. This 'normal' view of science neglects the important matter of all theories having limitations that flow from their premises.

A new way forward: complexity thinking

The role of philosophy in science relates to recent debate around the application of complexity theory to socio-ecological systems. The complexity sciences emerged in the 1960s as a field of study concerned with the emergence of structure, order and patterns from complex and apparently disordered (i.e. chaotic) systems (e.g. Prigogine and Stengers 1984). Complexity theories gained prominence in the 1990s with Kauffman's (1993) work on self-organising systems, and Holland's (1995) work on *complex adaptive systems*. In recent decades, complexity theory has been touted across a wide array of disciplines as a new and improved way of understanding the behaviour of complex systems. In environmental management, complexity theory is being used to devise better ways of modelling, predicting and managing the behaviour of complex socio-ecological systems, and dealing with the challenges posed by their inherent uncertainty (e.g. Espinosa and Walker 2011;

Norberg and Cumming 2013; Perez and Batten 2006). In organisational management, complexity is being used to find new ways of strategically manipulating organisations and their members to promote competition, adaptation and innovation (e.g. Ahrweiler 2013; Anderson 1999; Baumann and Siggelkow 2011; Buchanan 2011; March 1991). Complexity theory is also being applied in the knowledge management field, to help understand how knowledge is generated, shared and applied in organisations (e.g. Kakihara and Sorensen 2002; McElroy 2000; Stacey 2001).

Some scholars (e.g. Shackley *et al.* 1996; Stacey *et al.* 2000), however, criticise complexity-based approaches that attempt to improve prediction and control of system behaviour as simply being extensions of more traditional systems-based approaches. Complexity theory, along with quantum theory, presents a direct challenge to the Newtonian science-based understanding of reality that currently dominates western thinking, and has profound implications for how we understand the behaviour of living systems, both natural and human. Richardson (2008) argues that rather than thinking of complexity as a tool or method, managers should focus on understanding the *philosophy* of complexity, and the light it sheds on the limits of knowledge and how our underlying assumptions shape our worldviews and decisions. A significant barrier to natural scientists wishing to familiarise themselves with either philosophy or complexity is the language and terminology used in texts on these subjects, which are often impenetrable to those without formal training in the fields. Watzlawick *et al.* (2011) note that helping people reframe their problems requires translating potential solutions into their own language and presenting them in ways consistent with their conceptualising of reality (i.e. their paradigm).

It was with all this in mind that I set out to solve the riddle of knowledge transfer and adaptive management. What started as tentative probing into social science territory launched me on a neophyte's exploration through the history of science and philosophy, mathematical logic, quantum theory and the complexity sciences. *Complexity and Creative Capacity* traces this journey, and attempts to interpret some of the seemingly disparate ideas from these fields, and synthesise them into an alternative way of thinking about and framing our environmental problems. It does so by using complexity theory to surface and challenge the deeply held cultural assumptions that shape how we think about reality and knowledge, and to demonstrate how these assumptions are premised on a 'flattened' Newtonian view of a complex universe. In the chapters that follow, the underlying assumptions that shape our everyday understanding of cognition, learning, language, and social organisation are likewise exposed as Newtonian relics. On this basis, *Complexity and Creative Capacity* shows how our traditional approaches to generating and applying knowledge may be paradoxically exacerbating (or even generating) some of the wicked environmental problems we are attempting to solve. It implicates old constructs of knowledge transfer, adaptive management and adaptive capacity as potential culprits in a flawed problem formulation, which at best may be leading to ineffective environmental policy, and at worst the inadvertent creation of new problems.

Organisation of the book

Complexity and Creative Capacity is organised into seven chapters that attempt to build the reader's knowledge of the interdisciplinary subject matter in a step-by-step way. To prepare for the central themes that emerge in the later chapters, I highly recommend reading Edwin Abbott's (1884) delightful and widely available masterpiece *Flatland: A Romance of Many Dimensions*. By combining Victorian satire with mathematical fantasy, this slim volume ingeniously coaxes the layperson's mind into an understanding of the fourth dimension, and what it means to have a 'flattened' view of a complex universe.

Chapter 1 Introduction: the relationship between knowledge and adaptive capacity

The present chapter outlines the significance and purpose of the book and an overview of its content. It begins with a brief overview of the environmental management literature, outlining the role of *adaptive capacity*—the ability to learn and adapt to change—in maintaining the resilience of socio-ecological systems. Establishing tight feedback loops between social and ecological systems, as per adaptive management, is identified as a requisite for adaptive capacity, and knowledge transfer between science and management is identified as one of the key mechanisms for doing so. It is pointed out, however, that knowledge transfer efforts often fail to have the desired impact. A discussion follows on how knowledge transfer failures may partly result from those in the environmental sciences failing to appreciate contributions from the fields of organisational management, psychology, organisational learning, change management and knowledge management. It is pointed out that the ubiquity of failures across these disciplines suggests a common root to the knowledge transfer problem, one based in our shared cultural assumptions. Philosophy is posited as a means of uncovering cultural assumptions that lead to errors in problem formulation and inadvertently perpetuate or escalate the wicked environmental problems we are trying to solve. In particular, the philosophy of *complexity* is put forward as a means of shedding light on the limits of knowledge and how our underlying assumptions shape our worldviews and decisions.

Chapter 2 Unearthing assumptions: the co-evolution of Western science and philosophy

Chapter 2 presents a simplified overview of selected topics in the history of Western philosophy, science, and mathematics, to outline how the schools of thought relevant to our understanding of knowledge and reality evolved up to the post-war period. The discussion reveals the shifting and conflicting nature of the schools' underpinning assumptions, and illuminates the interrelationship between the evolution of philosophical ideas (such as atomism, reductionism, materialism, idealism, holism, process philosophy and postmodernism) and scientific ideas (such as classical mechanics, evolution, thermodynamics, organicism, quantum theory, cybernetics, chaos and complexity theory). The chapter concludes by suggesting that the competing themes that repeatedly emerge, subside and remerge in Western metaphysics aren't actually conflicting; rather, they are paradoxical elements of a singular, higher-order phenomenon. It closes by noting that in the chapters that follow mathematical logic will be used to 'prove' this paradoxical idea, and concepts from the complexity and chaos sciences will be used to help us understand how this is so.

Chapter 3 Unravelling paradox: mathematical logic and orders of change

This short chapter draws on examples from Watzlawick *et al.* (2011) and Bateson (2000) to introduce the reader to mathematical logic and how it can be applied to improve reasoning and resolve paradox. The discussion starts by using group theory to show how marked changes can take place *within* the parts of a group or system without effecting any change in the *overall* system, and how such internal change may paradoxically *reinforce* the pattern of the system whole instead. It then introduces the theory of logical types to show how system wholes are *qualitatively* (i.e. logically) different from their parts, the peculiar metamorphosis that occurs when one *jumps* from one logical level to another, and how paradox is generated when logical levels of phenomena are confused (i.e. system wholes are assumed to share the behaviours of system parts, and vice versa). This is followed by a discussion on how Gödel's theorem shows that logical systems always contain a 'hole', or a piece of missing information, which creates the requirement for a qualitative 'jump' when moving from one logical/system level to another. These theories are then related to Watzalwick *et al.*'s (2011) orders of change and principles of problem formulation. The chapter concludes by noting that the critical importance of these theories will become apparent in the exploration of the complexity sciences presented in the chapter that follows.

Chapter 4 Filling the 'hole': non-linearity, chaos and complex self-organising systems

This chapter begins with a discussion on 1920s metaphysical holism, which points to the tendency of wholes to be greater than the sum of their parts, and how holism was brought down in favour of reductionism in the 1930s because it was unable to answer the question 'what is it that is "more" in the whole?' This is followed by a discussion of the evolution of the complexity sciences, together with a step-by-step explanation of key conceptual contributions from the fields of non-linear mathematics, chaos theory, non-linear thermodynamics, autopoiesis and complexity science. The chapter concludes with a discussion on how the 'more' in the question posed by holism pertains to *emergent properties*—those properties that spontaneously self-organise in non-linear systems operating far from equilibrium, and the critical information missing from reductionist thinking, as per Gödel's incompleteness theorem. The chapter closes by noting that in furnishing this missing information, chaos and complexity theory have revealed holism and reductionism to be paradoxically wedded.

Chapter 5 Escaping Flatland: from a Newtonian to a complexity reality

This chapter weaves together the discussions from the previous chapters to challenge the underlying assumptions and philosophies of the Newtonian paradigm that currently dominates Western thinking and underpins our notions of knowledge and reality. It argues that the lens of our Newtonian assumptions has 'flattened' our view of a complex universe, and in doing so has created paradox in our understanding of it. A complexity-based paradigm is proposed as a more complete alternative to the Newtonian paradigm, one that unifies conflicting western philosophies such as reductionism and holism, and materialism and idealism. The proposed paradigm is presented as a synthesis of concepts from mathematical logic, relativity, quantum physics, holism, Whitehead's process philosophy, general systems theory, chaos theory, non-linear thermodynamics, autopoiesis and complexity theory. The chapter concludes by comparing and contrasting aspects of the complexity-based paradigm with that of the Newtonian paradigm.

Chapter 6 Through the complexity lens: cognition, learning, knowledge and organising

In this chapter key phenomena relevant to the problem of knowledge transfer are discussed: cognition, learning, knowledge and organising. They are described in relation to how they have been understood within the Newtonian paradigm, then compared with how they can be alternatively explained from the perspective of the proposed complexity-based paradigm, via notions such as embodied cognition, social construction of reality, and complex responsive processing. The discussion shows that while both paradigms clearly have predictive power when describing these phenomena, the complexity-based paradigm is able to resolve some of the paradoxes created by the core Newtonian assumption of a material, external and independent reality. The complexity paradigm is also shown to have profound implications for how we understand learning, knowledge, organising and other social phenomena.

Chapter 7 Implications: rethinking science, knowledge transfer and adaptive management

The chapter begins by outlining the limits of reductionist science revealed by complexity theory, and by comparing the role of science in the Newtonian versus the proposed complexity-based paradigm. Drawing on discussions from Chapter 6, Newtonian assumptions that inform our understanding of knowledge transfer, adaptive management, and adaptive capacity are shown as amounting to logical errors in problem formulation, as per Watzlawick *et al.* (2011). These errors are exposed as generators of paradoxical situations that perpetuate or escalate the environmental problems we are trying to solve, thereby inadvertently creating wicked problems. The chapter concludes with a reconceptualisation of the problem of knowledge transfer from a complexity perspective, proposing the concept of *creative capacity* as an alternative to *adaptive capacity* as a measure of resilience in socio-ecological systems. In doing so, knowledge transfer is revealed to be an ill-conceived strategy for enhancing system resilience and transcending wicked problems. The chapter closes by outlining a suite of possible strategies for building creative capacity.

Notes

1 Alice Calaprice, editor of *The New Quotable* Einstein (Einstein and Calaprice 2005), was unable to find the source of this quote, which may be a paraphrase of others.
2 See Chapter 3 for an explanation of the theorem. Franzen (2005) notes that Gödel's theorems have been widely invoked—often erroneously—in realms outside of mathematics and logic, in fields ranging from theology and literature to physics and postmodernism. These invocations are often inappropriate given the theorems only apply to formal theories of mathematical logic (specifically, consistent axiomatic systems capable of arithmetic). Where I relate Gödel's theorems to non-formal systems, it is as a *metaphoric* extrapolation of Gödel's notion "that the real behavior (truth) of nature is larger than any purely predicative, syntactic, model of it" (Maier 2006: 126). Where I relate Gödel to complexity in later chapters, however, note that Gödel's theorem is legitimately applied, as is discussed in the conclusion of Chapter 4.
3 Meaning a set of axioms or rules can never fully explain the system it is meant to account for.
4 Building on the ground-breaking work of Gregory Bateson (2000).
5 Chalmers (2002: 252) observes that although scientists are good at scientific progress, they are not "particularly adept at taking a step back from their work and describing or characterising the nature of that work".
6 Rovelli (2012) suggests the 'divorce' between science and philosophy worked for the last half of the twentieth century because the scientists operating in the 1930s were very, very smart, and most of our present-day science is founded on their ideas. He believes, however, that physics is at a point now where a return to philosophy is needed.
7 For this same reason, I was hesitant to risk making any reference to Sheldrake in this book.

References

Abbott, E. A. (1884). *Flatland: A Romance of Many Dimensions*. London: Seeley & Co. Retrieved from https://books.google.ca/books?id=139bAAAAQAAJ&dq=flatland+a+romance+of+many+dimensions&source=gbs_navlinks_s.

Ahrweiler, P. (2013). *Innovation in Complex Social Systems*. London: Routledge.

Anderson, P. (1999). Complexity theory and organization science. *Organization Science 10*(3), 216–232.

Argyris, C. and Schön, D. A. (1974). *Theory in Practice: Increasing Professional Effectiveness*. San Francisco, CA: Jossey-Bass.

Argyris, C. and Schön, D. A. (1978). *Organizational Learning: A Theory of Action Perspective*. Reading, MA: Addison-Wesley.

Argyris, C. and Schön, D. A. (1996). *Organizational Learning II: Theory, Methods and Practice*. Reading, MA: Addison-Wesley.

Arlettaz, R., Schaub, M., Fournier, J., Reichlin, T. S., Sierro, A., Watson, J. E. *et al.* (2010). From publications to public actions: When conservation biologists bridge the gap between research and implementation. *BioScience 60*(10), 835–842.

Armitage, D., Marschke, M. and Plummer, R. (2008). Adaptive co-management and the paradox of learning. *Global Environmental Change 18*(1), 86–98.

Bartunek, J. M. and Moch, M. K. (1987). First-order, second-order, and third-order change and organization development interventions: A cognitive approach. *Journal of Applied Behavioral Science 23*, 483–500.

Bateson, G. (2000). *Steps to an Ecology of Mind: Collected Essays in Anthropology, Psychiatry, Evolution, and Epistemology*. Chicago, IL: University of Chicago Press.

Bateson, G. (2002). *Mind and Nature: A Necessary Unity*. Cresskill, NJ: Hampton Press.

Baumann, O., & Siggelkow, N. (2011). Complexity and Competitive Advantage. In S. Macguire, P. Allen & B. McKelvey (Eds.), *The Sage Handbook of Complexity and Management* (pp. 494–505). London: Sage.

Bengtsson, J., Angelstam, P., Elmqvist, T., Emanuelsson, U., Folke, C., Ihse, M. *et al.* (2003). Reserves, resilience and dynamic landscapes. *Ambio 32*(6), 389–396.

Berkes, F. (2010). Devolution of environment and resources governance: Trends and future. *Environmental Conservation 37*(4), 489–500.

Berkes, F., Colding, J. and Folke, C. (2003). Introduction. In F. Berkes, J. Colding and C. Folke (eds), *Navigating Social-ecological Systems: Building Resilience for Complexity and Change* (pp. 1–29). Cambridge: Cambridge University Press.

Bohm, D. (2012). *Wholeness and the Implicate Order*. New York: Routledge.

Brown, V. A., Deane, P., Harris, J. A. and Russell, J. (2010). Towards a just and sustainable future. In V. Brown, J. Harris and J. Russell (Eds.), *Tackling Wicked Problems Through the Transdisciplinary Imagination* (pp. 3–15). London: Earthscan.

Buchanan, M. (2011). Power Laws and the New Science of Complexity Management. In B. McKelvey (Ed.), *Complexity: Critical Concepts* (pp. 2–10). London: Routledge.

Burnes, B. (2005). Complexity theories and organizational change. *International Journal of Management Reviews 7*(2), 73–90.

Carpenter, S. R. and Gunderson, L. H. (2001). Coping with collapse: Ecological and social dynamics in ecosystem management. *BioScience 51*(6), 451–457.

Carpenter, S. R., Walker, B. H., Anderies, J. M. and Abel, N. (2001). From metaphor to measurement: Resilience of what to what? *Ecosystems 4*, 765–771.

Carpenter, S. R., Brock, W. A. and Ludwig, D. (2002). Collapse, learning, and renewal. In L. H. Gunderson and C. S. Holling (eds), *Panarchy: Understanding Transformations in Human and Natural Systems* (pp. 173–193). Washington, DC: Island Press.

Cash, D. W., Clark, W. C., Alcock, F., Dickson, N. M., Eckley, N., Guston, D. H. *et al.* (2003). Knowledge systems for sustainable development. *Proceedings of the National Academy of Sciences 100*(14), 8086–8091.

Chalmers, A. (2002). *What Is This Thing Called Science?* St Lucia: University of Queensland Press.

Chapman, K. J., Horwitz, P., Scherrer, P. and Syme, G. (submitted). Knowledge that acts: A new participatory framework for closing the research-implementation gap. Unpublished manuscript submitted for journal publication in 2015.

Conklin, J., Basadur, M. and Van Patter, G. (2007). 'Rethinking wicked problems': Conversation between Jeff Conklin, Min Basadur and G.K. Van Patter. *NextD Journal 10*(Conversation 10.1), 30.

Curtin, C. G. (2014). Resilience design: Toward a synthesis of cognition, learning, and collaboration for adaptive problem solving in conservation and natural resource stewardship. *Ecology and Society 19*(2), 15.

Easterby-Smith, M., Lyles, M. A. and Tsang, E. W. K. (2008). Inter-organizational knowledge transfer: Current themes and future prospects. *Journal of Management Studies 45*(4), 677–690.

Ehrlich, P. R. and Kennedy, D. (2005). Millennium assessment of human behavior. *Science 309*(5734), 562–563.

Einstein, A. and Calaprice, A. (2005). *The New Quotable Einstein*. Princeton, NJ: Princeton University Press.

Espinosa, A. and Walker, J. (2011). *A Complexity Approach to Sustainability: Theory and Application* (vol. 1). London: Imperial College Press.

Festinger, L. (1957). *A Theory of Cognitive Dissonance*. Evanston, IL: Row Peterson.

Fischer, J., Dyball, R., Fazey, I., Gross, C., Dovers, S., Ehrlich, P. R. *et al*. (2012). Human behavior and sustainability. *Frontiers in Ecology and the Environment 10*(3), 153–160.

Folke, C., Colding, J. and Berkes, F. (2003). Synthesis: building resilience and adaptive capacity in social-ecological systems. In F. Berkes, J. Colding and C. Folke (eds), *Navigating Social-Ecological Systems: Building Resilience for Complexity and Change* (pp. 352–387). Cambridge: Cambridge University Press.

Folke, C., Hahn, T., Olsson, P. and Norberg, J. (2005). Adaptive governance of social-ecological systems. *Annual Review of Environment and Resources 30*, 441–474.

Franzen, T. (2005). *Gödel's Theorem: An Incomplete Guide to Its Use and Abuse*. Wellesley, MA: Peters.

Freeman, A. (2005). The sense of being glared at: What is it like to be a heretic? *Journal of Consciousness Studies 12*(6), 4–9.

Fulton, E. A., Jones, T., Boschetti, F., Chapman, K., Little, R. L., Syme, G. *et al*. (2013). Assessing the impact of stakeholder engagement in Management Strategy Evaluation. *International Journal of Economics and Management Engineering 3*(2), 82–98.

Funtowicz, S. O. and Ravetz, J. R. (1993). Science for the post-normal age. *Futures* (September), 739–755.

Gunderson, L. H. (2003). Adaptive dancing: Interactions between social resilience and ecological crises. In F. Berkes, J. Colding and C. Folke (eds), *Navigating Social-Ecological Systems: Building Resilience for Complexity and Change* (pp. 33–52). Cambridge: Cambridge University Press.

Gunderson, L. H. and Holling, C. S. (Eds.). (2002). *Panarchy: Understanding Transformations in Human and Natural Systems*. Washington, DC: Island Press.

Hedberg, B., Nystrom, P. C. and Starbuck, W. H. (1976). Camping on see-saws; Prescriptions for a self-designing organization. *Administrative Science Quarterly 21*(1), 41–65.

Heylighen, F., Cilliers, P. and Gershenson, C. (2007). Complexity and philosophy. In J. Bogg and R. Geyer (Eds.), *Complexity, Science and Society* (pp. 117–134). New York: Radcliffe Publishing.

Holland, J. H. (1995). *Hidden Order*. Reading, MA: Helix.

Holling, C. S. (1973). Resilience and stability of ecological systems. *Annual Review of Ecology and Systematics 1973*, 1–23.

Holling, C. S. (1978). *Adaptive Assessment and Management*. Chichester: Wiley.

Holling, C. S. and Meffe, K. (1996). Command and control and the pathology of natural resource management. *Conservation Biology 10*(2), 328–337.

Huber, G. P. (1991). Organizational learning: The contributing processes and the literature. *Organizational Science 2*, 88–115.

Huitema, D., Mostert, E., Egas, W., Moellenkamp, S., Pahl-Wostl, C. and Yalcin, R. (2009). Adaptive water governance: Assessing the institutional prescriptions of adaptive (co-)management from a governance perspective and defining a research agenda. *Ecology and Society 14*(1), 26.

Juarrero, A. (1999). *Dynamics in Action: Intentional Behaviour as a Complex System*. Cambridge, MA: MIT Press.

Kakihara, M. and Sorensen, C. (2002). Exploring knowledge emergence: From chaos to organizational knowledge. *Journal of Global Information Technology Management 5*(3), 48–66.

Kauffman, S. (1993). *The Origins of Order: Self-Organization and Selection in Evolution*. New York: Oxford University Press.

Kuhn, T. S. (1962). *The Structure of Scientific Revolutions*. Chicago, IL: University of Chicago Press.

Lyotard, J. F. (1984). *The Post-Modern Condition: A Report on Knowledge*. Minnesota, MN: University of Minnesota Press.

Maddox, J. (1981). A book for burning? *Nature 293*(24 September), 245–246.

Maddox, J. (1994). John Maddox on Sheldrake and book burning. Extract from BBC *Heretics of Science* documentary broadcast 19 July 1994, retrieved from www.youtube.com/watch?v=QcWOz1xjtsY (extract only).

Maier, J. (2006). Understanding the Complexity of Design. In D. Braha, A. Minai and Y. Bar-Yam (eds), *Complex Engineered Systems* (pp. 122–140). Berlin: Springer.

March, J. G. (1991). Exploration and exploitation in organizational learning. *Organization Science 2*(1), 71–87.

Markus, H. and Zajonc, R. B. (1985). The cognitive perspective in social psychology. In G. Lindzey and E. Aronson (eds), *The Handbook of Social Psychology* (vol. 1, pp. 137–230). Hillsdale, NJ: Lawrence Erlbaum.

McElroy, M. W. (2000). Integrating complexity theory, knowledge management and organizational learning. *Journal of Knowledge Management 4*(3), 195–203.

McKenzie-Mohr, D. (2011). *Fostering Sustainable Behavior: An Introduction to Community-Based Social Marketing*. Gabriola Island: New Society Publishers.

McNie, E. C. (2007). Reconciling the supply of scientific information with user demands: An analysis of the problem and review of the literature. *Environmental Science and Policy 10*(1), 17–38.

Medema, W., McIntosh, B. S. and Jeffrey, P. J. (2008). From premise to practice: A critical assessment of integrated water resources management and adaptive management approaches in the water sector. *Ecology and Society 13*(2), article 29.

Meppem, T. and Bourke, S. (1999). Different ways of knowing: a communicative turn toward sustainability. *Ecological Economics 30*(3), 389–404.

Michaels, S. (2009). Matching knowledge brokering strategies to environmental policy problems and settings. *Environmental Science and Policy 12*(7), 994.

Millennium Ecosystem Assessment. (2005). *Ecosystems and Human Well-being: Synthesis*. Washington, DC: Island Press.

Milton, J. and Boyd, J. R. (1867). *Paradise Lost*. New York: A.S. Barnes & Company.

National Research Council. (1999). *Global Environmental Change: Research Pathways for the Next Decade*. Washington, DC: National Academy Press.

Noë, A. and Thompson, E. (2004). Are there neural correlates of consciousness? *Journal of Consciousness Studies 11*(1), 3–28.

Nonaka, I. and Konno, N. (1998). The concept of 'Ba': Building a foundation for knowledge creation. *California Management Review 40*(3), 40–54.

Nonaka, I. and Takeuchi, H. (1995). *The Knowledge Creating Company: How Japanese Companies Create the Dynamics of Innovation*. New York: Oxford University Press.

Norberg, J. and Cumming, G. S. (2013). *Complexity Theory for a Sustainable Future*: New York: Columbia University Press.

Nyhan, B. and Reifler, J. (2010). When corrections fail: The persistence of political misperceptions. *Political Behavior 32*(2), 303–330.

Olsson, P., Folke, C. and Berkes, F. (2004). Adaptive comanagement for building resilience in social–ecological systems. *Environmental Management 34*(1), 75–90.

Palmer, J., Smith, T. and Mitchell, C. A. (2007). Creativity, ethics and transformation: key factors in a transdisciplinary application of systems methodology to resolving wicked problems in sustainability. Paper presented at the Systemic Development: Local Solutions in a Global Environment.

Peat, F. D. (1990). *Einstein's Moon: Bell's Theorem and the Curious Quest for Quantum Reality*. Chicago, IL: Contemporary Books.

Perez, P. and Batten, D. F. (2006). *Complex Science for a Complex World: Exploring Human Ecosystems with Agents*. Canberra: ANUE Press.

Peterson, G., Allen, C. R. and Holling, C. S. (1998). Ecological resilience, biodiversity and scale. *Ecosystems 1*, 6-18.

Plummer, R., Crona, B., Armitage, D. R., Olsson, P., Tengö, M. and Yudina, O. (2012). Adaptive comanagement: a systematic review and analysis. *Ecology and Society 17*(3), 11.

Prigogine, I. and Stengers, I. (1984). *Order Out of Chaos: Man's New Dialogue with Nature*. New York: Bantam Books.

Raymond, C. M., Fazey, I., Reed, M. S., Stringer, L. C., Robinson, G. M. and Evely, A. C. (2010). Integrating local and scientific knowledge for environmental management. *Journal of Environmental Management 91*(8), 1766–1777.

Resilience Alliance. (2015). Adaptive capacity. Retrieved 28 March 2015 from www.resalliance.org/index.php/adaptive_capacity.

Richardson, K. A. (2008). Managing complex organizations: Complexity thinking and the science and art of management. *Emergence 10*(2), 13–26.

Rittel, H. and Webber, M. (1973). Dilemmas in a general theory of planning. *Policy Sciences 4*, 155–169.

Roux, D. J., Rogers, K. H., Biggs, H. C., J., A. P. and Sergeant, A. (2006). Bridging the science–management divide: Moving from unidirectional knowledge transfer to knowledge interfacing and sharing. *Ecology and Society 11*(1), article 14.

Rovelli, C. (2012). Science is not about certainty: A philosophy of physics. *Edge*. Retrieved 15 September 2012 from https://edge.org/conversation/a-philosophy-of-physics.

Scheffer, M., Brock, W. A. and Westley, F. (2000). Mechanisms preventing optimum use of ecosystem services: An interdisciplinary theoretical analysis. *Ecosystems, 3*, 451–471.

Senge, P. M. (1990). *The Fifth Discipline: The Art and Practice of the Learning Organization*. New York: Currency Doubleday.

Senge, P. M., Kleiner, A., Roberts, C., Ross, R. B., Roth, G. and Smith, B. S. (1999). *The Dance of Change: the Challenge of Sustaining Momentum in Learning Organizations*. New York: Currency Doubleday.

Shackley, S., Wynne, B. and Waterton, C. (1996). Imagine complexity: The past, present and future potential of complex thinking. *Futures 28*(3), 201–225.

Sheldrake, R. (1981). *A New Science Of Life: The Hypothesis of Formative Causation*. Los Angeles, CA: J. P. Tarcher.

Shotter, J. (1994). *Conversational Realities: Constructing Life through Language*. London: Sage Publications.

Snowden, D. (2003). Complex acts of knowing: Paradox and descriptive self-awareness. *Bulletin of the American Society for Information Science and Technology 29*(4), 23–28.

Stacey, R. (2001). *Complex Responsive Processes in Organizations*. London: Routledge.

Stacey, R., Griffin, D. and Shaw, P. (2000). *Complexity and Management: Fad or Radical Challenge.* New York: Routledge.

Walker, B. H., Carpenter, S. R., Anderies, J. M., Abel, N., Cumming, G., Janssen, M. *et al.* (2002). Resilience management in social-ecological systems: a working hypothesis for a participatory approach. *Conservation Ecology 16*(1), article 14.

Walters, C. J. (1986). *Adaptive Management of Renewable Resources*. New York: McGraw-Hill.

Walters, C. J. (1997). Challenges in adaptive management of riparian and coastal ecosystems. *Conservation Ecology 1*(2), article 1.

Watzlawick, P., Weakland, J. H. and Fisch, R. (2011). *Change: Principles of Problem Formulation and Problem Resolution*. New York: Norton.

Wenger, E. (1998). *Communities of Practice: Learning, Meaning, and Identity*. New York: Cambridge University Press.

Wondolleck, J. M. and Yaffee, S. L. (2000). *Making Collaboration Work: Lessons from Innovation in Natural Resource Management*. Washington, DC: Island Press.

2 Unearthing assumptions

The co-evolution of Western science and philosophy

I fully agree with you about the significance and educational value of methodology as well as history and philosophy of science. So many people today—and even professional scientists—seem to me like someone who has seen thousands of trees but has never seen a forest. A knowledge of the historic and philosophical background gives that kind of independence from prejudices of his generation from which most scientists are suffering. This independence created by philosophical insight is—in my opinion—the mark of distinction between a mere artisan or specialist and a real seeker after truth.

(Albert Einstein to Robert A. Thornton, 7 December 1944)

Surfacing our assumptions

The first step in resolving intractable problems involves surfacing the deep assumptions that underlie our thinking around the problem at hand—in this case the problem of knowledge transfer and research application. This is the essential function of philosophy: the critical examination of the basis for fundamental beliefs and how they are expressed. The chief branches of philosophy include logic, metaphysics, epistemology, and ethics. Metaphysics is a branch of philosophy concerned with investigating the nature of reality and what is ultimately 'real' (Craig 1998). Metaphysical questions relate to the existence and nature of god, minds, bodies, time, space, causality, unity, identity, and the world in general. The central branch of metaphysics is ontology,[1] which examines questions of 'being' (i.e. existence) and what types of things exist (*ibid.*)—(i.e. what reality consists of). Epistemology is a branch of philosophy that "investigates how we can know and reason about that reality" (Heylighen *et al.* 2007). It asks: what is knowledge, how is it attained, and to what extent can a thing or a phenomenon be known?

Lorimar points out that despite modern science having evolved from a series of centuries old metaphysical and epistemological assumptions, today's scientists rarely concern themselves with epistemology or metaphysics. He underlines this with a quote[2] he attributes to the philosopher Willis Harman: "Epistemology is to scientists what ornithology is to birds!" But if one follows the logic of Einstein, Gödel and Watzlawick *et al.*, this is a perilous situation. Failure to recognise the core assumptions that frame how they think about the world renders scientists (and society in general) vulnerable to committing fundamental errors when attempting to formulate and resolve today's intractable and wicked problems. In the chapters that follow, these errors will be shown to lead to vicious circles and exacerbation of the very problems we are trying to solve. In the words of Einstein (1949: 684),[3] "Science without epistemology is—in so far as it is thinkable at all—primitive and muddled".

To surface my own underlying assumptions about reality and knowledge, and those which largely underpin science and the field of environmental management, I first needed to gain a basic understanding of the history of Western science, philosophy and mathematics. This chapter is a summary of my exploratory review of these topics. It attempts to reveal the shifting nature of their underpinning assumptions, and how they have entwined and mutually shaped each other since the time of the ancient Greeks. The chapter is, however, by no means scholarly or comprehensive. Drawing largely on the work of Capra (1997, 2010) and Heylighen *et al.* (2007), it is intended as a simple introduction for readers unfamiliar with the rich and extensively documented territory of Western science and philosophy. To assist, some key philosophical terms are simply defined in Table 2.1.

Ancient era and Middle Ages

Unity of opposites and atomism

In ancient times, people believed that the universe was held together by an uneasy tension between order and chaos. This tension is evident in Hesiod's *Theogony*, the Chinese notion of yin and yang, the terrible cross roads occupied by the Indian creator god Shiva, and the Pythagorean concepts of *peras* and *apeiron* (Briggs and Peat 1989). The fifth-century BC Pythagoreans were more interested in patterns than substance—for them, reality at its deepest level was mathematical (Bateson 2000). The Pythagoreans believed in a limitless void, without which 'things' could not be *distinguished*—successive numbers in a series, for instance (Kirk and Raven 1983: 340). In their view the cosmos was a "play of emptiness and form" (York 2013), of the limiters (*peras*) and unlimited (*apeiron*), which locked together to create an ordered universe (Kirk and Raven 1983).

The Pythagorean belief that 'things' are thrust into existence by the co-existence of their opposite 'nothing' (without which 'things' could not be distinguished) was also central to the ideas of Heraclitus, a pre-Socratic Greek philosopher. Heraclitus's *unity of opposites* posits that all existing entities consist of pairs of contrary properties. The opposition between these pairs is what brings the entities into being (see Graham 2011; Seibt 2013). Although different, these pairs are nonetheless interconnected. Graham (2011) illustrates this unity of opposites using the example of sleeping and waking. Neither concept can exist without the other, and they are connected as one necessarily changes around to the other. The wholeness resulting from this paradoxical relationship between connectedness and opposition was a central thread in Heraclitus's thinking. For example, he writes "Collections: wholes

Table 2.1 Definitions of selected philosophical terms

Term	*Definition*
Philosophy	The critical examination of the grounds for fundamental beliefs.
Metaphysics	Philosophic study concerned with determining the real nature of things, and questions about the existence and nature of minds, bodies, god, space, time, causality, unity, identity, and the world.
Ontology	The study of being, or what applies to everything that is real. It is concerned with identifying the kinds of things that actually exist.
Epistemology	The study of the nature, origin, and limits of human knowledge.
Logic	The study of the nature and types of logic, including the relation of logic to mathematics and other disciplines.
Materialism	The view that only physical things truly exist, that everything (including the human mind and will and the course of human history) are causally dependent upon physical processes, or even reducible to them. Sometimes also called *physicalism.*
Idealism	A view that stresses the central role of the ideal or the spiritual in the interpretation of experience. It holds that the world or reality exists essentially as spirit or consciousness, that abstractions and laws are more fundamental in reality than sensory things, or that whatever exists is known in dimensions that are chiefly mental—through thought and as ideas.
Dualism	The belief that mental things and physical things are fundamentally distinct kinds of entities. It implies that mind and body not only differ in meaning but also refer to different kinds of entities.
Panpsychism	A belief that everything in the world has some mental aspect. This view attributes some degree of consciousness—however small—even to apparently inert bits of matter.
A priori	Knowledge that is based on reason alone, independent of experience. An a priori argument reasons deductively from abstract general premises, e.g. 3 + 4 = 7, or all triangles have three sides.
A posteriori	Knowledge derived from experience, as opposed to a priori knowledge. An a posteriori argument relies upon specific information derived from the senses, e.g. Vancouver is on the coast of the Pacific Ocean.
Rationalism	The view that reason is the most reliable source and test of knowledge, that significant knowledge of the world can best be achieved by a priori means.
Empiricism	The view that experience is the most reliable source and test of knowledge, that significant knowledge about the world must be acquired by a posteriori means.
Positivism	Any knowledge system that confines itself to natural science and data derived from observation, and excludes a priori or metaphysical speculations.

Table 2.1 Continued

Term	*Definition*
Realism	The belief that material objects exist independently of our perception of them.
The Four Causes	Aristotle's four answers to the question of why something is, or how a thing comes about:
	Material cause The stuff from which the thing is made; what material is it made of? (e.g. clay).
	Formal cause The form or pattern of a thing; what 'type' of thing is it? (e.g. a head).
	Efficient cause Origin of a change or state of rest in something—the agent that imposed this form on that matter; what caused the thing to come into being? (e.g. the sculptor).
	Final cause The end or goal of a thing; what purpose is the thing meant to fulfil? (e.g. the production of a statue).
Determinism	The theory that all events, including moral choices, are completely determined by previous events or causes (i.e. efficient causes). It holds that the universe is utterly rational because complete knowledge of any given situation assures unerring knowledge of its future. Determinism is usually understood to preclude free will because it assumes that humans cannot act otherwise than they do.
Teleology	A philosophy whereby phenomena are understood to arise because of their purpose or design (i.e. they arise 'purposefully' towards their final intended form, rather than through efficient causes only). Also described as final causality. Natural teleology contends that natural entities possess intrinsic purpose which drives their function (e.g. an acorn grows into an oak tree because to be an oak tree is its purpose).
Reductionism	A view that asserts that entities are aggregates of smaller, more basic entities; that whole entities are equal to the sum of their parts, such that their behaviours can be explained by the behaviours of their parts.
Holism	A view that asserts that the universe, and particularly nature, is made up of interacting wholes that are more than the sum of their parts; i.e. the behaviour of an entity's parts cannot fully explain the behaviour of the whole entity.
Atomism	Any doctrine that explains complex phenomena in terms of aggregates of fixed particles or units.
Upward causation	The behaviour of the parts determining the behaviour of the whole (e.g. as per reductionism, determinism and Aristotle's efficient causes).
Downward causation	The behaviour of the whole determining or constraining the behaviour of the parts (e.g. as per holism, teleology and Aristotle's final causes).

Sources: Encyclopaedia Britannica (www.britannica.com); Garth Kemerling (www.philosophypages.com).

and not wholes; brought together, pulled apart; sung in unison, sung in conflict; from all things one and from one all things" (cited in Graham 2011). Heraclitus also believed that reality (i.e. the universe) is in continuous flux and that all things flow. This is evident in his exhortation that you can never "step twice into the same river" (cited in Kirk and Raven 1983: 214). For Heraclitus, reality was a constellation of *processes* rather than *things:* "the river is not an object, but an ever-changing flow; the sun is not a thing, but a flaming fire" (Rescher 2008). As such, Heraclitus's philosophy centred on the never-ending process of 'becoming' rather than on 'being' (Graham 2012). This was a philosophy of change and process rather than substance.

Parmenides, however, disputed these ideas. He argued that there is no such state as 'non-being', and hence there can be no void. Instead, he believed that all of reality is comprised of a single, indivisible mass, and that change and motion are illusory, there being no void to move into (Berryman 2011). Thus, there is only 'being', one kind of 'thing' (Silverman 2008). In an effort to reconcile the differences between Heraclitus's ever-changing reality and Parmenides's unchanging, singular reality, Leucippus and Democritus proposed their materialist theory of *atomism*[4] (Berryman 2011). Atomism is a *reductionist* philosophy. It posits that all entities are comprised of small indivisible particles known as atoms, separated from each other by infinite void. Atoms thus form a constellation of unchanging Parmenidian entities floating around in a Pythagorean void. When objects change their shapes this is the result of atoms shifting their configuration (*ibid.*). In this view, reality is both unchanging (in the form of irreducible atoms), and always changing (in the form of shifting configurations of atoms). The atomists were *mechanistic* and *deterministic* in their belief that all phenomena (i.e. configurations of atoms) could be understood by exploring their prior circumstances (i.e. the prior interactions that caused a particular atomic configuration to materialise; *ibid.*). This is consistent with Aristotle's material and efficient causes, which will be discussed in the next section, and with the notion of *upward causation* described in Table 2.1. The atomists also believed the world had no purpose, divine intervention or *final cause.*

Socratic order

With the coming of the Socratic era (400–300 BC), notions of an ever-changing universe, continuously brought into being by the interplay of chaos and order, gave way to a focus on the ordered aspects of the universe (Briggs and Peat 1989). The universe was now seen as 'being' rather than 'becoming'. Plato resisted atomism's simple bottom-up explanation of the cosmos, arguing that mechanistic collisions of atoms couldn't possibly be responsible for the 'ideal' beauty of the physical world (see Furley 1999). Aristotle, Plato's pupil, viewed matter as being continuous. He refuted the idea of a void, believing it to be in violation of physical principles (Sorensen 2012). Like Plato, Aristotle also believed the atomists' deterministic account of the physical world was insufficient. From his perspective, truly knowing a thing required more than taking it apart and tracing the motions of its various ingredients: it also required knowing something as its "present whole" (Furley 1999: 25). To this end, Aristotle formulated his theory of *four causes* (Table 2.1) to explain why something is or how it comes about (see Falcon 2012). A thing's *material cause* relates to what it's made of, or its matter. A thing's *formal cause* pertains to its form, structure or pattern. *Efficient cause* is the origin of change in a thing's state. And lastly, a thing's *final cause* is its end goal or purpose, which is related to the notion *downward causation* described in Table 2.1. Plato and Aristotle both rejected Democritus's reduction of all things to purely material and efficient causes. They believed that natural phenomena (and the wider universe in Plato's case) had purpose, or final causes, and that enquiries about the world should prioritise teleological investigation into what purpose a phenomena serves (Zeyl 2014). As such, the atomist view of reality eventually fell away in favour of the Aristotelian, until it was resurrected hundreds of years later, by seventeenth-century Renaissance thinkers (van Melsen 2015).

Socratic philosophers were concerned with the role of matter (material causes) and form (formal causes) in comprising reality. Individual things were seen to have *properties*, such as colour, size, shape, location, and so on. Plato believed that the properties of things exist independently of those things, as abstract, eternal forms or ideas, created by an intelligent designer in a realm beyond space and time (Mcintosh 2012). As such, all things in the world unfold in accordance to these ideal forms. Plato's ideal forms are an early expression of *idealism*—the belief that reality is fundamentally ideal, abstract or spiritual. Aristotle disagreed with Plato's idea of an abstract, intelligently designed world of forms, believing instead that properties dwelled *within* material things as innate qualities (Cohen 2009). He believed that forms could not exist separately to substance. For example, although he believed the soul and the body to be distinct, he believed they were inseparable (i.e. the soul cannot exist without the body; Shields 2011). According to Aristotle, these indwelling properties are in essence final causes, and all things in the world unfold in accordance with their 'internal natures' (e.g. a seed has a tree as its final cause, therefore it becomes a tree; Cohen 2009). According to Seibt (2013), "The history of Western metaphysics since Aristotle was mainly focused on elaborating various versions of substance metaphysics—in fact, as one might say with Hegel, 'the history of [Western] metaphysics is the tendency towards substance'".

Metaphysically, Aristotle's ideas follow the middle ground between materialism and dualism (Shields 2011). In terms of epistemology, Aristotle was both an empiricist, in that he believed that experience (a posteriori knowledge) was the best means of knowing efficient and material causes, and an idealist, in that he believed intuition and reason (a priori knowledge) were needed to gain true knowledge about the 'essence' of things (Pajares 2012). Aristotle viewed the world as an ordered place, with all beings ordered in increasingly complex hierarchies (the Great Chain of Being). He believed in the human capacity to understand such order (Ruse 2013). These, and other aspects of Aristotle's philosophy and science came to largely dominate Western thinking for the next two thousand years, with an authority that Capra (1997) describes as almost as unquestioned as that of the church.

With respect to cosmology, some Greek thinkers, such as Aristarchus, believed the earth revolved around the sun. Aristotle, on the other hand, believed that the earth was the centre of the universe, and that celestial bodies, including the sun, moved in perfect circles around it. Actual observations of planetary movements did not, however, conform with Aristotle's cosmology, primarily because the outer planets appeared to move backward in a retrograde motion before moving forward again. In the second century BC, Ptolemy reconciled these observations with Aristotle's geocentric (earth-centred) universe by developing the idea that the planets traced epicycles[5] while orbiting the earth. Ptolemy's convoluted system of epicycles enabled close approximate predictions of planetary movements while maintaining the earth at the centre of solar system (Rabin 2010). This geocentric view of the cosmos was preserved as an article of faith in medieval Europe, remaining largely unchallenged for the next 1600 years.

After the collapse of the Greek–Roman empires, Western civilisation descended into the anarchy and turmoil of the Dark Ages. However, in the twelfth century, the work of ancient Greek philosophers was rediscovered and resurrected by medieval universities. Efforts by medieval 'schoolmen' to reconcile Greek philosophy—primarily that of Aristotle—with Christian theology led to the emergence of *scholasticism* (see Tamayao 2012). The scholastics revered reason, and as such scholasticism was more a method of critical thought than a philosophy. They worked to synthesise Greek logic, science, pagan philosophy and Christian theology, with a focus on Socratic *dialectal reasoning*. This is a method of resolving argument through a process of presenting arguments (theses) and counter-arguments (antitheses), leading to a synthesis of the two arguments, or some form of improvement in the argument (Cirne-Lima 1997). Scholasticism was exemplified by thinkers such as Thomas Aquinas and William Ockham, and became the dominant philosophy of the Middle Ages, reigning in Western universities until 1500 AD.

Scientific revolution

Empiricism and reductionism

In the early 1500s, the Polish polymath Nicolaus Copernicus devised his heliocentric (sun-centred) theory of the solar system. His theory was able to mathematically explain the retrograde motions of the planets without using Ptolemy's epicycles, and satisfied celestial observations with fewer explanations (Rabin 2010). Although fearful of how his controversial ideas would be received, Copernicus finally published his theory while on his deathbed. By placing man outside the centre of the universe, Copernicus shook the foundations of his society and triggered the Scientific Revolution.

The scientific revolution (i.e. early modern period) in the seventeenth and eighteenth centuries was the next major shift in Western thought. This period saw the rise in *empiricism* (the belief that knowledge is primarily derived from the senses and experience) in the works of Bacon, Locke and Hume. Galileo removed quality from science by restricting science to phenomena that could be measured and quantified (Rogers 2004). This in turn led to the formulation of the scientific method and Comte's doctrine of *positivism* (Bourdeau 2011). Positivism contends that empirical science, with its reliance on a posteriori knowledge, is the best approach for explaining how the world works. Empiricism presented a challenge to the dominant rationalist and idealist epistemologies of the time, which deemed the human senses unreliable and privileged thought, intuition and reason (a priori knowledge) over observation (Prinz 2006). Using telescopic observations, Galileo found and published empirical evidence that supported and fine-tuned Copernicus's heliocentric celestial system. For his effort, Galileo was deemed a heretic and imprisoned for the remainder of his life.

It was during this time that the French philosopher and mathematician René Descartes unified algebra and geometry to invent analytical geometry and Cartesian coordinates. He also devised a revolutionary concept of analytic thinking called *reductionism*. Reductionism is premised on the idea that to understand a complex phenomenon, you simply have to pull it apart and analyse its component parts. Descartes's formulation of reductionism coincided with the invention of the clock, and his reductionist worldview is widely compared to that of a *clockwork universe* (Heylighen *et al.* 2007). Descartes did not believe in the 'void' and as such, held that all phenomena could be infinitely subdivided, there being no elementary particles, as believed by the ancient atomists (see Hatfield 2008). Nor did he believe in final causes. Instead, he believed all phenomena come into being mechanistically via cause and effect (i.e. through efficient causes, or upward causation), while also being subject to nature's basic and immutable mathematical laws. He believed these laws to be determined by God, implying a mechanical and divinely determined universe (Bertrand 2010; Psillos 2004).

While Descartes viewed the universe, animals and the human body as complicated clockwork machines (Capra 1997), this view conflicted with prevailing religious notions of humans having free will over their actions. Descartes proposed a *dualist* philosophy of mind and matter to resolve the contradiction, proposing that the human mind is of a different category of substance than that of the body, the 'ghost in the machine' so to speak, and therefore not subject to deterministic laws (Heylighen *et al.* 2007; Robinson 2011). This created philosophy's tenacious 'mind–body' problem, which continues to create debate to this day (Hatfield 2008). Descartes's dualist notion of reality was then countered by *monist* philosophers contending that reality consists of only one substance: either mind or matter. This heralded philosophy's modern materialist–idealist split, with materialists (e.g. Hobbes) contending that mind and matter are in fact both matter, and idealists (e.g. Berkeley) contending that mind and matter are both products of the mind (deVries 2004; Mastin 2008).

Materialism and classical mechanics

Although rejected by Descartes, the ancient metaphysics of atomism were revived by the likes of Galileo, Gassendi, Cordemoy, Bacon, Hobbes and Newton. These great seventeenth and eighteenth century thinkers believed in the material existence of elementary particles, and argued that Descartes's view would mean there is no corporeal substance to the universe (Smith 2010). They also discarded Plato and Aristotle's final causes in favour of the ancient atomistic view of a deterministic universe, shaped through cause and effect (efficient and material causes); albeit one whereby particles and the forces between them are created by God (Capra 2010).

In 1687 Sir Isaac Newton laid the foundation of classical (or Newtonian) mechanics, with his laws of motion and theory of universal gravity (see Smith 2007). Since antiquity, time and space had both been regarded as relative, abstract concepts, with space being a measurement of the relative arrangements of different material bodies, and time a measure of the relative cycles of change, or motion (Rynasiewicz 2004). Newton, countered this by introducing the concept of *absolute* space and time, whereby space and time are regarded as real and independent aspects of reality (*ibid.*). In this view, time exists independently of the observer and proceeds at its own pace through the universe. Likewise, space is also external and independent of observation. Thus, in classical mechanics change is simply a spatial rearrangement of material components set against (and independent of) the backdrop of absolute space and time, their movement governed by deterministic laws of cause and effect (i.e. *upward* causation). By knowing the initial positions and velocities of particles or components that make up a system, theoretically one can predict exactly where a system will be in a given moment in time (i.e. the system's evolution can be predicted). In theory, one can also reverse the sequence of events to deduce exactly where a system *was* at a given moment in the past (i.e. the so-called 'the arrow of time' is theoretically *reversible* in deterministic systems). Like Descartes's clockwork universe, the linearity of Newton's theories reflects a mechanistic conception of a simple, predictable cosmos (Capra 2010). Newton's absolutist theory of space and time was a matter of much controversy among *relationists* of the time, among them the German polymath Gottfried Leibniz. Lieibniz argued that space made no sense except when viewed in the relative location of two bodies, and time made no sense except in relation to the movement of bodies (Ferraro 2007). This controversy died away with the success of Newton's ideas, until it resurged much later in the nineteenth century.

Historically, mathematicians had been plagued by their inability to devise an equation whereby the exact speed of an accelerating or decelerating body could be determined at a given time. Newton revolutionised mathematics by devising differential calculus[6] as a means of solving the problem (Capra 1997). Briggs and Peat (1989) describe how Newton's laws of motion and gravitation are based on *linear* differential equations (equations having solutions which can be *added* together to create new solutions). Linear equations do not incorporate feedback, meaning A may cause B, but B has no effect on A. In such equations, small changes cause small effects, and large changes either cause large effects or effects that can be explained by summing up many small changes. Solutions for one equation can be generalised to others. For example, using Newton's law of universal gravitation, total gravitational force can be calculated by adding up the gravitational forces contributed by different masses. The basic properties of the solar system could now be predicted. Other scientists quickly came to rely on Newton's differential equations to explain a wide variety of phenomena, including plant growth, the motion of fluids, and burning coal (*ibid.*).

The Cartesian and Newtonian ontology of the universe can be summed up as one of *materialism* and *determinism*.[7] Materialism dictates that reality (including all phenomena, physical, biological, mental and social) ultimately consists of matter. Consciousness, life, organisation and purpose are reduced to epiphenomena arising from configurations of particles of matter in time and space (Heylighen *et al.* 2007). According to Heylighen *et al.* (*ibid.*), this worldview assumes that all phenomena, including ourselves and our minds, are equal to the sum of their parts. 'Being' is simply the static condition of an object located in space. A system evolves as its elementary particles move about. However, "there is no way for particles to merge, divide, appear or disappear … everything that exists now has existed from the beginning of time and will continue to exist, albeit in a somewhat different configuration" (Heylighen *et al.* 2007: 119).

Once the initial conditions of the universe have been established, everything that follows is the inevitable result of the domino-cascade of cause and effect, every event the result of a continuous chain of prior events (i.e. via *upward causation,* or *efficient causes*). With complete and accurate information, the future of any part of a system can be predicted with total confidence. Philosophers call this understanding of how the world works *physical determinism*.[8] In such a world, there is no free will, no true novelty or creation (see Prigogine and Stengers 1984), no chaos, mystery or miracles, just predictable rearrangement of matter through cause and effect. This view of a world without uncertainty was famously described by the eighteenth-century French mathematician Pierre Laplace:

> An intellect which at a certain moment would know all forces that set nature in motion, and all positions of all items of which nature is composed, if this intellect were also vast enough to submit these data to analysis, it would embrace in a single formula the movements of the greatest bodies of the universe and those of the tiniest atom; for such an intellect nothing would be uncertain and the future just like the past would be present before its eyes.
>
> (Laplace 1951: 4)

As such, the spirit of the scientific revolution was the belief that by using empirical methods and applying Descartes's coordinates and Newton's mechanical laws and differential equations, everything in the universe could be reduced to a point where it could be explained and predicted (Briggs and Peat 1989). This way of thinking, which is now often referred to as the *Newtonian paradigm*, would dominate science for the next 200 years, driven by the material-technological concerns of a rising merchant capitalism, and its need for economic production and military development (Foster and Clark 2008, citing others).

Romantic and Industrial eras

Idealism and panpsychism

Philosophers such as Hobbes, Gassendi, and Marx embodied the ethos of the scientific era by embracing materialist philosophies and rejecting the notion of a separate spirit or mind substance.[9] The way materialists think about the world is predicated (very generally) on two key assumptions:

- The exterior world is real, independent of us, objective and knowable (the materialist metaphysics/ontology).
- Knowledge is created as we observe the world and our senses send data about it to our minds (the materialist epistemology, otherwise understood as *realism*; Ridling 2001; Ritchie and Lewis 2003).

In the eighteenth and nineteenth centuries, however, idealist philosophies grew strength as followers of the Romantic Movement revolted against the industrial revolution, rejected notions of a purely mechanistic universe, and embraced the natural world (Barry 1999). Philosophers of the time, such as Kant, were also attempting to bridge the epistemological divide between rationalism and empiricism created during the scientific revolution (McQuillan 2012). In contrast to materialism, idealist philosophies[10] prioritise the mental, spiritual or abstract nature of reality over the physical. There are several variations of idealism, differing in the relative importance they give the material world. The more extreme varieties posit that space and time are purely mental constructs. For example, *solipsism* postulates that the only thing sure to exist is one's own mind (see Solomon and Higgins 2004). Idealism rejects the notion of a strictly deterministic universe shaped purely by cause and effect, asserting that mind, or consciousness, is able to exert free will and shape reality (see Becker 2004) through a process of downward causation. Very generally, the way strict idealists think about the world is predicated on two key assumptions:

- The existence of the external world is dubious and subjective (the idealist metaphysics/ontology).
- Knowledge is subjective and inside our minds; we can only apprehend the world as sensations and ideas in our minds so we cannot truly know it (the idealist epistemology; Solomon and Higgins 2004).

Berkeley, for example, was a proponent of an extreme form of idealism known as subjective idealism, which denies the existence of any material substance and argues that objects only exist in the minds of observers (deVries 2004). In the eighteenth century, Kant set off the 'Copernican revolution of philosophy' by merging scholasticism, empiricism and rationalism to devise *transcendental idealism* (also known as empirical realism), which dominated philosophy into the nineteenth century (McCormick 2005). Transcendental idealism accepts the existence of an independent material reality, but argues that we can never truly know it because our mind shapes our experience of reality. Later in the nineteenth century, philosophy in Germany and Britain came to be dominated by Hegel's *absolute idealism*, which posits that objects must exist in our consciousness (via a thought by God) before they can become real and material (McQuillan 2012).

Other monist philosophers, such as Spinoza and Leibniz in the seventeenth century, adopted the idea of *panpsychism*, which straddles the idealist-materialist divide by claiming that everything has a mind *and* a soul, therefore all things are both mind and matter (Skrbina 2007). Similarly, in the eighteenth century, La Mettrie and Diderot advanced *vitalistic materialism*, which discarded the notion of 'soul' in favour of the idea that a mind-like nature was present in all matter (*ibid.*). Panpsychism also developed rapidly among nineteenth-century German philosophers, such as Schopenhauer and Goethe, who argued against the division of mind and matter: "no matter without mind, no mind without matter" (*ibid.*).

Capra (1997) describes how the German idealists and panpsychists of the eighteenth and nineteenth centuries also revived ancient *holist* traditions focusing on nature in its organic form, in opposition to the Cartesian view of the world as a machine. For example, Kant believed that science could only offer mechanical explanations that could not be used to understand life, which he saw as self-organising wholes. Likewise, Goethe viewed nature as a great, singular, harmonious whole. Hegel revived the ancient works of Heraclitus, and as such, saw the universe as 'becoming' in dialectical process of being and non-being (deVries 2004). He also rejected reductionism and atomism, believing objects had universal qualities that could not be reduced to a set of properties, and that objects should be treated as primary wholes (Stern 1990). Hegel saw dialectical reasoning[11] (the iterative progression from thesis, to antithesis, then synthesis) as a means of progressing toward absolute *truth*, which he viewed holistically as *completeness*, rather than as an assertion about something (Stokes 2012). While Aristotle believed a thesis and its opposing antithesis could not both be true, Hegel viewed them as partial truths, brought together into a higher truth by synthesis. For example, instead of seeing falsified scientific theories as completely wrong, Hegel saw them as partial conceptions of a *whole* truth (*ibid.*).

Vitalism

During this time a complementary thread of thinking was being revived by nineteenth-century biologists: the ancient doctrine of *vitalism*. According to Hattiangadi (1994), the deepening ontological divide between eighteenth century natural and moral philosophy had resulted in complete separation in the nineteenth century:

> The word 'philosopher' came to be reserved for an apologist for idealism or perhaps an opponent of thoroughgoing mechanism in any form, and the expression 'scientist' came into use for the thoroughgoing mechanist who followed the experimental method of investigation. These two professions came to inhabit different parts of the university, and came to adopt different curricula, and thus and only so were able to keep the peace.
>
> (Hattiangadi 1994: 231)

At this time a debate erupted among scientists about how to explain, in a material universe governed by mechanical laws, the form and purpose of living things, as well as movement perception, development and life (Bechtel and Richardson 1998). On one side of the debate were those who attempted to explain these things in purely mechanistic terms, as products of matter acted upon by electrical, chemical and magnetic forces (i.e. substance monism). On the other side were those attempted to explain life's form and purpose by adding to the equation a special non-material life force (i.e. substance dualism)—these were the *vitalists* (Hattiangadi 1994). Vitalist biologists opposed reductionism, believing that chemical and physical laws could not on their own explain 'life' (Capra 1997). In contrast to Descartes's notion that animals and human bodies are mechanical 'automata', the vitalists believed some underlying force—perhaps in the form of spirit, fluid, or organising principle—was responsible for giving bodies life (Bechtel and Richardson 1998). As such, from the vitalist perspective "living matter is greater than the sum of its parts because of some life force" (Gilbert and Sarkar 2000: 1).

Among the vitalists were microbiologist Louis Pasteur, embryologist Hans Driesch and anatomist Xaviar Bichat (Bechtel and Richardson 1998). Driesch, for example, used the term *entelechy* to describe a vitalist force that organises physico-chemical processes into self-regulating and goal seeking organisms (Weber and Esfeld 2003). By the early twentieth century, a number of prominent physicists, including Niels Bohr and Erwin Schrödinger, had also taken up the vitalist belief that organisms are subject to unknown physical laws, unlike those which govern inert matter (Mayr 1988). An important aspect of the vitalist doctrine was the belief that living things have a purpose, or a final goal. As such, living things originate with an enfolded blueprint that causes them to develop, or unfold, into what they were intended (Bechtel and Richardson 1998). In other words, their parts self-organise into a pre-ordained whole. This idea, known as *formative teleology*,[12] has its origins in Aristotle's final causes. According to Stacey *et al.* (2000), formative teleology was a central feature of Kant's idealist belief that living things develop with purposeful intelligence, that their purpose is to reach their final, predetermined forms.[13] Just as deterministic notions of a mechanistic universe governed by cause and effect eliminate the possibility of true novelty emerging, so does formative teleology (Stacey *et al.* 2000).

Evolution

During the second half of the nineteenth century the microscope was invented and mechanistic views came back into favour as major advances were made in biology and the other sciences (Capra 1997). In physics, empirical evidence had come to show that magnetic, electrical and caloric (heat) forces were interchangeable and unified as energy, undermining the vitalist idea of a life-giving force unique to living things (Hattiangadi 1994). Advances in physiology were also presenting mechanical explanations for previously mysterious life processes. Perhaps the most decisive challenge to vitalism, however, arrived in 1859 against the backdrop of British colonial exploration, when Darwin (1870) and Wallace (1871) proposed their theory of evolution. The theory was premised on the belief that all species have diverged from common ancestors through a process of *natural selection*. Natural selection acts on chance variation; organisms having traits that enable them to 'fit' their environment and survive competition will pass these traits on to next generations. As species evolve and diverge through the *survival of the fittest*, natural systems become more complex and ordered over time. Since Aristotle, Western society had largely believed in final causes, that there was a cosmic force in the world driving all things to greater perfection (Mayr 1999). Darwin, however, did not believe in the notion of 'soul', and his theory denied the role of any cosmic teleology.[14,15] He saw evolution as purposeless with no end in mind (Mayr 1999). By arguing that there is no grand design in nature, Darwin had dissolved the mind-body problem and the need to explain form and purpose in living things (Hattiangadi 1994). By the late nineteenth century, support for vitalism had dwindled, and it has since been discredited.

Darwin's theories came to have enormous impact on modern thought, shaping science as well as the Western outlook and way of life into the twentieth century (Mayr 1999). For example, the term 'survival of the fittest' led to social Darwinism, which used evolutionary theory to justify social progress of superior groups (Barry 1999). This resonated with Britain's colonial aspirations and later with fascism and Nazism in Europe. Western culture also became embedded with belief that competition is necessary for the creation of 'healthy' systems (*ibid.*). In contrast, Russian naturalists and evolutionary theorists were unconvinced by Darwin's emphasis on natural selection. They believed cooperation (or *mutual aid*), not competition, to be the basis of survival and evolution. Among the Russians it was commonly argued that Darwin and Wallace had been overtly influenced by dominant socio-economic and political theory emerging from an overpopulated, viciously capitalist England[16]—a country starkly juxtaposed to Russia's sparsely inhabited steppes and communal peasant life (Todes 2009).

Marx and Engels likewise argued that Darwin's theories of natural selection and competition anthropomorphised nature, having been derived from bourgeois *social* theories, as admitted by Darwin himself (Foster and Clark 2008). These social theories included Malthus's 1798 *Principle of Population*, Hobbes's 1651 doctrine of 'war of everyone against everyone', and Adam Smith's 1776 theories on economic competition and division of labour. Engels saw the social application of Darwin's ideas as a case *of double transference*—transferring social theories about society onto nature, then applying them back to society as natural laws—with distorted and unbalanced interpretations of social phenomena being the result (Foster and Clark 2008). In 1859, Karl Marx introduced his philosophy of *historical materialism* (later modified into *dialectical materialism*) in *A Contribution to the Critique of Political Economy*. Historical materialism proposes that individual and societal interests are reconciled through a process of progressive dialectical change (see Albrecht 2005). Allen suggests that the closely timed publication of Darwin's *Origin of Species*, Marx's *Critique of Political Economy* and Virchow's essay 'The mechanistic conception of life' together codified "an explicitly materialistic interpretation of the natural and social worlds" (Allen 2005: 269).

Thermodynamics

Another major scientific advance of this era was that of classical thermodynamics. Prigogine and Stengers (1984) describe its origination as follows. From the 1700s, scientists and engineers had been puzzling over why it was not possible to invent a perpetual motion machine. Whenever energy was fed into a machine, despite recycling efforts, some of it was always mysteriously lost, requiring continuous inputs of replacement energy. In 1824, while working on steam engines, the French physicist Sadi Carnot realised that mechanical energy was not actually lost; rather it was dissipated as heat, which is disordered energy the system cannot use again. This understanding was formulated into the second law of thermodynamics, which states that closed systems (systems with no external inputs of matter or energy) always tend to move from order to disorder, with disorder steadily increasing over time. Physicists coined the word *entropy* to describe the disorder in a system. Entropy introduced the *arrow of time* into science by showing that closed systems proceed in a particular direction (from order to disorder), which cannot be reversed.

Thus, classical (or linear) thermodynamics suggested a world becoming increasingly disordered, where "the entire world-machine is running down and will eventually grind to a halt" (Capra 1997: 98). This presented a significant challenge to the Newtonian assumption that all phenomena in the universe are mechanically reversible. In the 1870s, as an attempt to reconcile Newton's laws with entropy, Ludwig Boltzmann, a Viennese physicist, devised *probability theory*, otherwise known as statistical mechanics. According to Briggs and Peat (1989), probability theory suggests that the movements of the trillions of atoms comprising matter are so complicated that over time it becomes increasingly unlikely that they will stay in ordered relationships—that is, they trend towards disorder, or entropy, instead. Eventually the entire universe will dissolve into random disorder (i.e. chaos) and become an inert, homogeneous soup. From Boltzmann's theory it was believed that science would eventually understand the underlying mechanism of entropy, and by minimising or circumventing it, learn to control and predict the behaviour of complicated systems (*ibid.*).

Machine age

Organicism and ecology

In the early twentieth century a number of biologists formed a new anti-mechanist/reductionist movement which was also anti-vitalist: *organicism*, a term coined by William Ritter in 1919. Organicism grew out of opposition to the "'Machine model' of life and mind" and the mechanised rending of life, labour and culture caused by industrialisation, urbanisation, the First World War and Darwinism (Harrington 1996: xvii). Among organicism's chief adherents were the embryologists Ross Harrison, Paul Weiss, and Joseph Needham; as well as philosophers of biology E. S. Russell and J. H. Woodger. According to Gilbert and Sarkar (2000), between the first and second world wars organicism was the framework for embryology in both the USA and Europe. Organicism was also strongly influenced by emergentist philosophers such as C. D. Broad and Lloyd Morgan, and holist process philosophers such as Alfred Whitehead and Jan Smuts (Denton *et al.* 2013).

Like the vitalists, organicists believed that living system wholes (such as cells, tissues, organs, organisms, ecosystems) are greater than the sum of their parts, and cannot be understood solely by looking at their components. Unlike vitalists, however, organicists were materialists, who did not believe that vital, non-material forces were responsible for organising the patterns they observed in living material. As such, organicist biologists transcended mechanistic notions in their belief that patterns of *organising relations* were the key to understanding biological form, while also believing "no separate, nonphysical entity is required for understanding life" (i.e. they rejected both Cartesian reductionism and vitalism; Capra 1997: 25). They were particularly interested in multi-levelled structures of systems within systems (*ibid.*). A central principle of organicism is that each level of complexity (e.g. tissue) in a complex system arises from the reciprocal *interactions* of its parts (e.g. cells), and that its properties emerge from those interactions, rather than simply being a summation of the parts themselves (Gilbert and Sarkar 2000). For example, water has a property of 'wetness' that emerges from the interactions of its molecules, but those individual molecules taken alone are not 'wet', nor can the property of 'wetness' be predicted by examining them in isolation (*ibid.*). Another central principle of organicism is that each level of organisation is governed by its own set of rules and principles (*ibid.*). Therefore, biology cannot be reduced to laws of physics and chemistry. Whereas Darwinian biologists saw natural selection and adaptation as the forces of evolution and the source of life's complexity, the organicists saw evolution as largely driven by life's intrinsic tendency toward increasing complexity and organisation, and emergent innovation at the metabolic level (Deichmann and Travis 2010).

Aristotle and Plato had regarded living things as having 'essences' that gave them their form, thereby rendering them as static, unchanging entities (McDonough 2010). As such, prior to the introduction of Darwin's theory of evolution, nature studies were typically descriptive natural histories (e.g. the works of von Humbolt, Larmarck and Hutton). The term *ecology* was first defined in 1866 by Haeckel, a Darwinian scientist, as the study of *relationships* between an organism and its natural surroundings. However, ecology was not taken up as distinct discipline until the first decades of the twentieth century, when the work of ecologists came to parallel that of organicist biologists by focusing on the "irreducible wholeness" of plant and animal communities (Capra 1997: 32). In its early days the field was dominated by the American botanist Frederick Clements. In 1905 he introduced his concept of the vegetation community,[17] which behaves as a kind of organism, what ecologists would later refer to as a *super-organism*. He proposed that these vegetation communities develop into predetermined and harmonised climax communities through a process of *succession*. The South African grassland ecologist, John Phillips (1931) later built on the work of Clements and Smuts[18] to propose his concept of the *biotic community* a community comprised of both plants and animals operating holistically as a complex organism.

General relativity

The early twentieth century also saw a momentous holist challenge mounted against Newton's classical physics. In Newton's theories, space and time are regarded as opposing, independent and unrelated concepts. They are also absolute, in that they appear the same to all observers (see Rynasiewicz 2004). However, in 1916, Albert Einstein transcended Newton's theories when he expanded his special theory of relativity into the *general theory of relativity,* which includes gravity as a property of space and time. This theory is premised on the notion that space and time are unified in the fourth dimension (known mathematically as Minkowski space) as a singular entity known as *spacetime*. By uniting the two on a 'higher plane' or dimension, Einstein's theory echoes Heraclitus's 2000-year-old concept of the 'unity of opposites', whereby all phenomena consist of pairs of opposing properties (Capra 2010). What appear to be separate unrelated or opposing phenomena can be seen to be parts of whole when put into the context of a higher dimension.[19] Capra (*ibid.*) uses the image of a donut ring sliced by a plane (Figure 2.1) as a metaphor to demonstrate this concept. When viewed in two dimensions the two circles (A and B) appear to be unrelated entities. However, when viewed from the higher order 3D perspective, the circles are revealed as being part of the same entity—a donut (C).

Newton's law of gravitation describes gravity as the attractive forces between massive objects; he saw gravity as operating instantaneously through empty space, but was unable to explain how it transmitted its forces across the vacuum (Huggett and Hoefer 2009). In Einstein's general relativity theory, gravity is not a 'force'; rather it is the result of distortions, or curves, in spacetime. As described by Overduin (2007), Einstein's theory depicts spacetime as a two dimensional surface contorted (like a rubber sheet) into dips and valleys by the massive objects occupying space. Spacetime is curved by matter, producing what we experience as gravity, and conversely, spacetime's curvature guides how matter moves. Measurements of space and time are relative and dependent on the observer's state of motion. Because the structure of spacetime is dictated by matter, the notion of empty space becomes meaningless (Capra 2010). As such, spacetime is not a fixed background, as per Newton's absolute notion of space and time; rather it is a dynamical structure that interacts with matter (Wuthrich 2010). Thus, while existing absolutely, spacetime behaves *relationally* (Overduin 2007). Einstein's general theory of relativity went on to become one of modern physics' essential tools. It has since had an impressive track record in terms of accurately predicting a wide range of phenomena, including black holes (where the orderly laws of physics break down), background microwave radiation, and the bending of light around planets (Wudka 2006). Although still an active field, it is now regarded as incomplete because it fails to account for *quantum effects* (Rickles 2008).

Emergentism, holism and process philosophy

Einstein's general theory of relativity bolstered a number of new philosophical challenges to reductionism, particularly by *emergentist* philosophers. In 1920, the

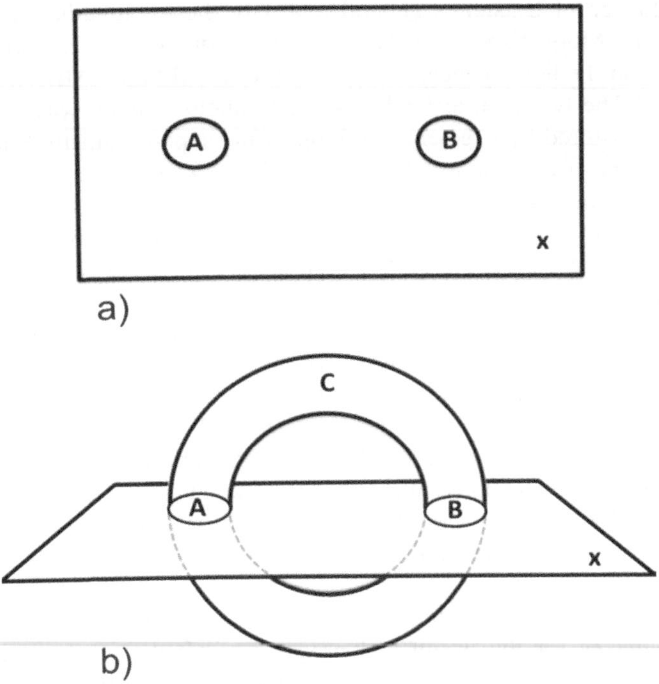

Figure 2.1 A donut ring incised by a plane, viewed first in two dimensions (a) then in three (b), depicting how two seemingly unrelated phenomena are revealed as parts of a whole when put into the context of a higher dimension

Source: adapted from Capra (2010).

Australian philosopher Samuel Alexander published his magnum opus *Space, Time and Deity*, which presented a new metaphysical system with unified spacetime as the fundamental entity of the universe. In Alexander's view, all existing things are continuously connected in spacetime through groupings of motions, and organised into a hierarchy of existence whereby each layer emerges from the next (Thomas 2012). Alexander's system[20] was influenced by panpsychic philosophies of Kant and Spinoza, the idealist philosophies of Plato and Hegel, Einstein and Minkowski's theories of spacetime, and mathematical logic (Thomas 2012). In 1923 the animal psychologist Lloyd Morgan published *Emergent Evolution* (1923), where he used the terms *emergence* and *emergent property* to describe the appearance of novel products of evolution, which could not be predicted or explained by combinations of prior properties (Weber and Esfeld 2003). Morgan's ideas were further developed by the English scientist, mathematician and philosopher C. D. Broad, in his treatise *The Mind and Its Place in Nature* (1925). Broad used the term emergent property to describe the unique properties that emerge at higher levels of complexity, properties which cannot be inferred from the nature of their parts (Gustavsson 2010).

The influence of Einstein's general relativity theory, together with that of Alexander and Morgan's work, is also seen in the *process philosophy* of Alfred Whitehead and the holist concepts of the controversial South African statesman Jan Smuts.[21] The two men arrived at similar conclusions by way of different approaches. Inspired by Clements's notion of the superorganism, Smuts fused Darwin's theory of evolution with Einstein's theory of relativity (Moller 2006). From this fusion he devised the term *holism* in his 1927 book *Holism and Evolution* (Smuts 1927), defining it as nature's fundamental tendency to dynamically form wholes that are greater than the sum of their parts. In his view, wholes have their own causal force, which they exert downward onto their parts (which have their own upward causal force, from which the whole emerges). This enables the wholes to be self-reproducing. Smuts foreshadowed the findings of complexity scientists later in the century with his notions that (i) the whole and its parts reciprocally influence and determine each other, and (ii) holism is a self-reproducing evolutionary force. Smuts believed Darwin overemphasised the importance of natural selection, which "powerfully reinforced and exaggerated the mechanistic conception of the universe" (*ibid*.: 197). In response, he extended Darwin's theory with his holistic theory of evolution:

> [Holism] underlies the synthetic tendency in the universe, and is the principle which makes for the origin and progress of wholes in the universe … Evolution is nothing but the gradual development and stratification of progressive series of wholes, stretching from the inorganic beginnings to the highest levels of spiritual creation.
>
> (Smuts 1927: ix)

Whitehead was a highly regarded mathematician who had published the groundbreaking *Principia Mathematica* (1913) with Bertrand Russell. Whitehead believed it was the West's ontology of substances that was causing mathematicians and physicists to struggle, as they attempted to treat as substantial those phenomena better understood as *process* (Malone-France 2007). This belief led him to write the seminal *Process and Reality* in 1929 (Sherburne 1981), which built on Hegel's ideas and disputed Newton's cosmology of bits of matter floating around in absolute space and time. Whitehead's process philosophy revives Heraclitus's notions of the 'continuous flow' of reality, and of reality as a 'process of becoming' through tension between opposites. It is also related to the dialectic of 'being' and 'not being' elaborated by Hegel in the nineteenth century. The foundational argument of process philosophy is that reality is better understood in terms of *processes* than as *things*, and that reality is premised on change (Rescher 2008). This is in stark opposition to the idea that reality is primarily made up 'things' (with change being secondary or accidental) that has biased Western metaphysics and science since the time of the early atomists and Aristotle (*ibid*.).

Whitehead premised his theory on Einstein's notion of unified spacetime[22] (Nixon 2010). He believed that the "physical world is bound together by a general type of relatedness which constitutes it into an extensive continuum," a spatial–

temporal extensiveness that forms a "vast nexus extending beyond our immediate epoch" (Sherburne 1981: 81). In process philosophy, reality has no material foundation; what we perceive as substantial entities, or matter, are actually just composites of indefinite "fluctuating processes organized into stable structures" (Rescher 2008). Whitehead uses the term *actual occasions* to describe the smallest units of process that comprise reality, ephemeral "drops of experience, complex and interdependent" (Sherburne 1981: 205). Griffin (1988: 151) defines an actual occasion as "a momentary experiential event, each of which occupies (or constitutes) a region that is spatial as well as temporal". These 'actual occasions' interact in a continuous process of bringing forth the future, so at each moment "the many enter into complex unity" (Sherburne 1981: 34). Thus, the universe is continuously *emerging* or *becoming,* in what Whitehead calls a "creative advance into novelty" (*ibid.*: 33). As such novelty is not just possible, it is endlessly emerging: the universe is not deterministic, and the arrow of time cannot be reversed.

Whitehead (Sherburne 1981: 29) also believed that Aristotle had "led to a wild overstressing of the notion of final causes during the Christian age; and thence, by a reaction, to the correlative overstressing of the notion of 'efficient causes' during the modern scientific period". As such, Whitehead proposes a compromise between final and efficient causality: while each emerging moment is novel, it is simultaneously constrained by the past. The configuration of actual occasions that comprise a given moment is *somewhat* consequential on every moment that preceded it (i.e. the result of cause and effect). In other words, while the self-creating unity of the universe (the final cause) ensures every moment is new and different from the previous, cause and effect (efficient causes) results in each moment *typically* being similar to that which preceded it (*ibid.*). Seemingly enduring 'things' are actually just stable patterns in time. Time gives immaterial reality the *illusion* of substance.

In addition to providing a synthesis between final (downward) and efficient (upward) causality in the universe, Whitehead's process philosophy is considered a bridge between materialist and idealist philosophies, and a development in the panpsychic ideas of the previous century (Skrbina 2007). Whitehead did not believe in the division of mind and matter. While he describes God as the primordial basis of all actual occasions, God himself is nonetheless an actual occasion, as is the remainder of the universe (Sherburne 1981).

Quantum mechanics

Interestingly, some of the central themes of process philosophy and holism—such as the insubstantial nature of reality and importance of the relations between 'things' rather than the things themselves—were simultaneously emerging in physics as scientists began probing the atomic and subatomic worlds. Kumar (2009) provides a rich account of this history, which launched in 1897 with J. J. Thomson's discovery of the electron. This sparked scientific realisation that atoms, once thought to be the smallest entities in the universe, might be comprised of much smaller *subatomic particles*. In 1900, the German physicist Max Planck discovered that energy is emitted in tiny packets, what he called *quanta*. This was contrary to the orthodoxy of the day, which held that energy travels in waves. Planck's discovery caught the attention of a young patent office clerk in Switzerland named Albert Einstein. Inspired, Einstein devised his 1905 theory proposing that light is quantacised into tiny particles called *photons*. Experimentation soon revealed the validity of both the wave and quantum theories of light—wave theory was needed to explain light diffraction and interference and quantum theory was needed to understand the photoelectric effect (the observation that metals emit electrons when light is shone on them).

In 1913, the young Danish physicist Niels Bohr developed a quantacised model of the atom. He proposed that as an atom becomes more or less energised, its electrons jump between orbits, emitting or absorbing discrete quanta of energy as they do so. Bizarrely, when committing a *quantum jump* an electron *does not* trace a trajectory. It simply occupies a certain orbit in one moment and another orbit in the next, *without* traversing the space between. Another puzzling discovery was made by Louis-Victor de Broglie in 1923. He found that matter at the subatomic level, like energy, also has co-existing wave and particle-like properties, what he called *wave–particle duality*. In 1926, Weiner Heisenberg and Max Born introduced *matrix mechanics*, a formulation for explaining quantum phenomena, and Erwin Schrödinger introduced his equivalent *wave mechanics*. Wave mechanics is centred on describing a subatomic entity's (electron, proton, photon, etc.) *wave function*—a mathematical representation of everything known about the entity's properties (i.e. position, momentum, energy, etc.)—and how it changes over time. These and other quantum-scale theories and discoveries were assembled into a cohesive theoretical and mathematical framework called *quantum mechanics*. Although experiments had proven quantum mechanics to consistently make accurate predictions about the atomic and subatomic (i.e. quantum) world, the question of what quantum mechanics 'said' about the nature of reality remained (Kumar 2009). In late 1927, the top minds in physics, including Bohr, Heisenberg, Einstein, Born, Schrödinger and Planck,[23] assembled over two successive conferences to debate this question. The dominant view that emerged was the *Copenhagen interpretation* of quantum mechanics, based on (i) Born's probabilistic interpretation of the wave function, (ii) Heisenberg's uncertainty principle, (iii) Bohr's principle of complementarity and (iv) wave function collapse (Kumar 2009).

Max Born's (1926) probabilistic interpretation of the wave function states that Schrödinger's wave function can only ever calculate the *probability* of finding a particle in given place. This was a direct challenge to Newtonian determinism, which assumes that precise predictions can be made about any phenomena given complete information. Born changed the frame in which quantum waves were understood, from that of a three-dimensional wave (like a sound wave or water wave) to that of a probability wave representing the abstract probabilities of *all* the particle's possible states. Quantum-scale particles were now understood to exist in a world of abstract probabilities. Paradoxically, this meant that particles could never be said to *exist* or *not exist* in a particular place; nor do could they be said to change their positions or remain at rest (Capra 2010). The only thing that changes is the probability of a particle existing in a particular location.

Newtonian determinism was further undermined by Heisenberg's (1927) uncertainty principle. It states that two properties of a sub-atomic particle cannot be simultaneously measured, because the more precisely you measure one property (e.g. momentum, energy, or position), the more uncertain other quantities become. Bohr's (1928) principle of complementarity similarly states that the wave and particle aspects of energy and matter are complementary but exclusive (like two sides of the same coin), as only one aspect can be measured at a time. Matter and energy present as either waves or particles, depending on how an experiment is set up. Bohr furthered this argument by postulating that the act of observing or measuring a particle causes its wave function (encompassing all its possible states) to collapse into a single state, which becomes observed reality. These principles were used to explain the results of the famously weird double-slit experiment,[24] in which photons exhibit wave or particle behaviour depending on *how they are observed.*

The Copenhagen interpretation posits that subatomic entities do not possess any absolute properties on their own. Rather, their properties arise from the interaction between the observer and the observed entity. This means the quantum world does not become real until it is observed or measured—a notion that strangely echoes Kant's transcendental idealism. At the quantum scale, the idea of a scientist playing the role of detached observer is shattered, along with the Newtonian assumption of an objective reality (Wheeler 1973).

In showing that apparently solid material objects may dissolve into "wave-like patterns of probabilities" at the subatomic level (Capra 1997: 30), quantum physics also served a heavy blow to the materialist assumptions of the Newtonian paradigm. Although molecules and atoms had been found to consist of subatomic particles, these particles are not isolated things in the Newtonian–atomic sense. Rather they are interconnections between things, a "complicated web of relations between various parts of a unified whole", which in turn are interconnections between other things, and so on (Capra 2010: 138). Instead of ordinary processes (like windstorms) being caused by very small *things* (like atoms), it was actually the combining of very small *processes* that was producing ordinary things (like a table, or a planet) (Rescher 2008). In keeping with Whitehead's process philosophy, it appeared that there might be no 'things' in the end, only interconnecting processes. Descartes's belief that all phenomena could be infinitely subdivided, that there are no indivisible pieces of matter in the universe, now seemed a more accurate view of reality than that of Newton and other atomists. In the words of Niels Bohr (1958: 57), "Isolated material particles are abstractions, their properties being definable and observable only through their interactions with other systems". Henry Stapp elaborates on this by noting "An elementary particle is not an independently existing unanalyzable entity. It is, in essence, a set of relationships that reaches outward to other things" (quoted in Capra 2010: 139).

By showing mind and matter to be of similar nature and part of a single system, quantum physics was also dismantling the philosophical dualism that had separated mind and universe since Descartes (Jeans 1934). Materialist notions of a universe filled with objects existing independently of the human mind gave way to idealism among many early-twentieth-century physicists, who thought perhaps there could in fact be a 'ghost in the machine'—some form of universal consciousness helping to shape reality, perhaps even God. For example, the physicist Sir James Jeans wrote:

> the stream of knowledge is heading towards a non-mechanical reality; the Universe begins to look more like a great thought than like a great machine. Mind no longer appears as an accidental intruder into the realm of matter; we are beginning to suspect that we ought rather to hail it as the creator and governor of the realm of matter—not of course our individual minds, but the mind in which the atoms out of which our individual minds have grown exist as thoughts.
>
> (Jeans 1930: 137)

This view—that the fundamental physical components of the universe have mental properties—signalled a return to the monist notions espoused in Spinoza's panpsychism and a radical departure from the dominant mechanistic conception of the universe. More recently, the American quantum physicist David Bohm proposed that consciousness is deeply rooted in the universe's 'implicate' order, and therefore present to some degree in all material forms, as a "rudimentary mind-like behaviour of matter" (Bohm and Hiley 1993). D'Espagnat (1979) likewise states that experimental results from quantum physics dispute the notion of a world existing independently of human consciousness. As such, Skrbina (2007) argues that panpsychism is no longer in question, rather it is a matter of what form it takes.

Not everyone, however, was in agreement with the Copenhagen interpretation's ontological drift toward idealism. Kumar (2009) describes how Einstein, for example, was outraged. While he agreed that quantum mechanics was mathematically correct and logically consistent, Einstein could not accept the Copenhagen interpretation of what quantum mechanics say about the nature of reality. He believed the interpretation was incomplete, that there must be hidden variables at play. Once revealed, these variables would explain the bizarre behaviour of quantum systems, allowing quantum mechanics to be reformulated into a deterministic theory that accommodates an observer-independent quantum reality. In 1935, Einstein, Podolsky and Rosen presented their so-called EPR paper. In it they introduced a thought experiment where two previously interacting (entangled) particles, A and B, separate such that they are now light years apart. They proved that by measuring the momentum and position of particle A one could simultaneously derive the position and momentum of particle B without ever directly measuring or observing it. This posed a problem for the Copenhagen interpretation, which postulates that particles do not become real or have actual properties until they are measured. For this aspect of the interpretation to hold up, particle A must instantaneously transmit information about its measured values to particle B. Einstein argued that no reasonable definition of reality would permit this mysterious faster-than-the-speed-of-light transmission, and that the locality of cause-and-effect must be preserved. He declared the Copenhagen interpretation incomplete, contending that there must be local hidden variables at play. Bohr, however weakly, successfully countered by arguing that 'spooky action at a distance' (as Einstein called it) does exist, that *non-locality* is an intrinsic feature of the quantum realm. As such the observer dependence of reality is maintained. Bohr and Einstein had reached an impasse, which they battled over until their deaths decades later.

Eintsein and his views became increasingly marginalised. This was partly due to the failure of viable alternatives to emerge, and partly due to Bohr and Heisenberg's dogmatic promotion of the Copenhagen interpretation.. The Copenhagen interpretation of quantum mechanics dominated for the next 60 years, becoming the most successful theory of all time and the basis of most modern technology (Kumar 2009). Later experiments supported Bohr's allowance for non-local faster-than-light influences between previously entangled particles (see Aspect 2007), including *quantum teleportation*[25] (Bouwmeester *et al.* 1997). *Non-separability* now appears to be a central feature of reality—particles that were once involved somehow remain part of a single system (Gribbin 1984).[26] Because every particle in the universe has been involved with other particles going back to the Big Bang, some quantum physicists now believe that non-separability is a demonstration of ontological holism (Gribben 1984). In the view of David Bohm and Basil Hiley (1993), for instance, the universe is an undivided whole, and quantum mechanics is better described by *organicism* than by mechanism (Musser 2013).

Logical positivism

Despite its obvious compatibilities with the new quantum science, idealism's sway over Western metaphysics waned in the 1920s, when it was overtaken by materialist philosophers aiming to return science to its reductionist and empirical roots (Uebel 2011). Vienna was the scene of much intellectual ferment in the 1920s. A varied group of scientists and philosophers, led by Moritz Schlick and calling themselves the Vienna Circle, met weekly to discuss problems in the philosophy of science and mathematics. Prominent among them were the logician Rudolph Carnap, physicist Philipp Frank, mathematician Kurt Gödel and the social scientist Otto Neurath. The circle became very interested in the seminal work of Austrian philosopher Ludwig Wittgenstein: *Tractatus Logicio-Philosophicus* (1922). Wittgenstein was largely influenced by his conversations with the mathematician Bertrand Russell about the foundations of logic and philosophy (Biletzki and Matar 2009). In the *Tractatus*, Wittgenstein applies modern logic to reveal the relationship between language and reality. Biletzki (2009) outlines how Wittgenstein was particularly concerned about the distinction between things that can be expressed using language (i.e. thought) versus those which can only be *shown*. He felt this was the principal problem of philosophy. Wittgenstein believed that the underlying logic structure of language limits what can meaningfully be said and thought about the world, particularly in relation to language itself and to metaphysics. His main premise was that the structure of language is determined by the structure of reality. Thoughts, and therefore language, are accurate 'pictures', or models, of the world; anything that cannot be 'pictured', is therefore beyond language, and is ascribed as 'nonsense'. Thus, mystical philosophical concerns, such as metaphysics, aesthetics and ethics, though meaningful, cannot be *discussed* in a way that is not 'nonsense'. Wittgenstein believed that the limits of language and thought should be the limits of philosophy. Thus, philosophy should be discarded for logical practices that reflect the 'real' world (i.e. that which can be 'pictured'), because attempts to discuss anything external to the real world (i.e. the mystical world) result only in nonsense (Biletzki and Matar 2009).

Uebel (2011) describes how the Circle, inspired by Wittgenstein's insights about logical truths and seeming opposition to metaphysics (Wittgenstein himself said he was misinterpreted by the Circle), took a strong empiricist and positivist stance and tasked themselves with "clearing the stables of academic philosophy". Rationalist and idealist epistemologies (e.g. those of Descartes, Leibniz, Spinoza and Kant) premised on the notion that knowledge can be gained a priori through intuition, thought and reason, were bitterly opposed. For the Circle, the only meaningful a priori knowledge was derived through *analysis*, formal proof using logic or mathematics. As such, the Circle relegated the role of philosophy to that of clarifying assertions and problems though formal logical analysis (analytic philosophy). Metaphysics, theology and ethics were viewed as meaningless nonsense and discarded altogether. Synthetic empirical statements were only meaningful if they could be somehow tested in the field. So it was that the Vienna Circle expanded the domain of empiricism and shrunk the role of philosophy in science. In doing so they brought forward an extreme form of empiricism called *logical positivism,* which was premised on the notion that statements were only true or meaningful if they could be verified by observation (a posteriori). They advocated *induction*-based approaches to generating theory, a posteriori from facts. In this view, science progresses by generating facts through experiment and observation, and then logically deriving theories that fit the facts through a process of induction (Uebel 2011). They also introduced the philosophy of *physicalism*, a unifying belief that all phenomena can be reduced to physics-based explanations, or that all that exists is physical (Shanker 1996b).

Demise of organicism

As holistic notions revolutionised 1920s thinking in the fields of biology and ecology, a countermovement spawned by the logical positivists was afoot. The positivists' concept of physicalism conflicted with one of central themes of organicism: that the rules of biology are not reducible to that of physics. A new crop of young biologists were now embracing mechanical physicalism as a means of 'raising' the credibility of biology to that of the 'hard' physical sciences, while rejecting the application of any 'soft' holistic concepts to the field (Allen 2005). Their thinking was in line with the Vienna Circle's manifesto, which equated the overthrow of organicist ideas in biology with its mission of ridding metaphysics from science and philosophy (Shanker 1996a). It was around this time that ecology began to part ways with idealism and organicism. The split widened when Alfred Volterra and Vito Lotka first introduced the principles of thermodynamics to explain predator–prey interactions in 1926 (Keller and Golley 2000). In 1927, Elton presented the concept of 'food cycles' (a precursor to 'food webs') and the idea that feeding relationships were the central organising feature of ecosystems (Elton 1927). Around the same time, Henry Gleason (1926) introduced population theory, whereby species distribution is a product of chance and individuals reacting to environmental conditions, rather than self-organising communities. Since its introduction, Clements's notion of the superorganism had been hotly debated by those who rejected his attribution of "mysterious organising properties" (Levins and Lewontin 1985: 135) to plant communities as idealist and teleological (Foster and Clark 2008). The debate raged until 1935, when Arthur Tansley published *The Use and Abuse of Vegetational Concepts and Terms* (1935). In his article, Tansley attacks Smuts's holism and the organicist premises of Clements and Phillips's superorganisms, climax communities and biotic communities. He also rejects the idea of emergent wholes not being predictable from the sum of their parts, suggesting that any unpredictability is simply due to the observer's lack of knowledge about the component properties. Tansley instead proposed a reductionist approach to ecology, centred on the concept of *ecosystem*,[27] meaning a physical system that is:

- composed of both organic and inorganic components tending toward stable dynamic equilibrium through their interactions; and
- part of a hierarchy of systems, from the universe down to the atom.

From this point on, trophic dynamics—the flow of energy and food through ecosystems—became the dominant focus of ecological studies. Smuts and Phillips's efforts to use holistic concepts to justify racist policies in South Africa (Foster and Clark 2008), the Nazis' adoption of an anti-Semitic holism, and the Communists' uptake of holism in their philosophy of dialectical materialism, served to further discredit holistic thinking in both ecology and biology (Gilbert and Sarkar 2000).

System theory

Tansley's ecosystem concept emerged just as *systems science* began to develop. Ludwig von Bertalanffy was a biologist who was part of the Vienna Circle during the 1920s. Although educated in the neo-positivism of the group, his unorthodox interests, including German mysticism, "precluded his becoming a good positivist" (Bertalanffy 1969: 12). Bertalanffy was influenced by organicist biologists, process philosophers, holists, ecologists and quantum physicists of the early twentieth century. He was concerned that mechanistic approaches to science neglected or denied the essential phenomena of life, and believed they should be replaced with a holistic approach founded in biology rather than physics. Bertalanffy advocated the organicist understanding of the organism as a whole, or a system, and believed the main objective of biology should be discovering the organising principles of the organism's various levels (*ibid.*).

In the 1930s, Bertalanffy combined the systems-thinking ideas of the time to devise a formal theory of living systems (*ibid.*). He was familiar with the mathematical and mechanical models used to describe physical systems, but observed that the systems they described were closed and near equilibrium—relatively static compositions of molecules and matter that eventually succumbed to entropy. At this time, biology had been making major inroads in the areas of homeostasis and metabolism (e.g. Cannon 1932), building on the earlier work of nineteenth-century physiologist Claude Bernard. Bertalanffy realised that living systems are open systems operating far from equilibrium, using their metabolisms to 'feed' on a continuous "Heraclitan flow" of matter and energy from their environments as a means of maintaining their steady state, i.e. their aliveness (Bertalanffy 1969: 160). Open systems, unlike closed systems, can become *more* ordered with time, causing entropy to decrease. This view of living systems compares to vortices in a stream: stable, ordered structures far from equilibrium, formed by the constant through-flow of water. If the water stills, the vortices collapse. Likewise, without a constant through-flow of matter and energy, the second law of thermodynamics 'kicks in' and living systems die and decompose into disorder. Thus, Bertalanffy's theory of living systems reconciles the opposing notions of a world becoming increasingly simple and disordered via the second law of thermodynamics, and a living world becoming increasingly complex and ordered via emergent evolution.

With the passing of the Aristotelian conception of a static world, Bertalanffy believed *dynamic interaction* to be the central problem for all fields in modern science. In 1945, he introduced general system theory (Bertalanffy 1969) with the aim of unifying the various fragmented natural and social sciences. This is a theory of 'wholeness', comprised of a doctrine of principles that apply to *all* systems, be they physical, biological or social, and irrespective of the elements and forces and involved. It is premised on understanding systems as sets of elements standing in interrelation among themselves and with environment—tight relations between elements give the whole coherence, identity and autonomy (Heylighen *et al.* 2007). Heylighen *et al.* (*ibid.*, citing others) summarise some of system theory's key concepts as follows. A system is separated from its environment by a *boundary*, which gives it *identity*, and across which matter, energy and information are exchanged as *inputs* and *outputs*. Systems become coupled together when outputs of one system are used as inputs by another, creating feedback loops. When multiple systems couple, they form a *network.* Once a network reaches sufficient *coherence* and begins to exhibit *emergent properties* it becomes a *supersystem.* Systems and supersystems are organised into *hierarchies,* extending upward to larger wholes, and downward to smaller subsystems. The theory has a *black box* view of subsystems: they are seen as abstract relationships and the *processes* they perform within the larger whole, rather than as independent material components. The structure of subsystems is considered irrelevant to their function, e.g. neurons and transistor chips perform the same information processing function, and are considered *isomorphisms*. Thus, system theory looks at the *patterns* of organisation that are common to all phenomena, rather than a phenomenon's material parts. Systems theory applies reductionism when looking downward at subsystems (i.e. upward causation), and emergence and holism when looking upward at the system whole (downward causation); both are equally important. Systems coupled into supersystems are constrained by the downward causation exerted by the whole, meaning they have to obey the rules of relationship that maintain the coherence (i.e. emergent properties) of the supersystem and can no longer act independently.

With his general system theory Bertalanffy proposed that many biological, behavioural and social phenomena could be explained by common mathematical models that transcended physics to address the problems of order, organisation, wholeness, and teleology—problems otherwise alien to mechanistic science. Although this did not happen until 30 years later, when Ilya Prigogine pioneered the field of non-linear thermodynamics (see Chapter 4), Bertalanffy's work formed a major scientific movement nonetheless, which, together with cybernetics, spawned a number of new fields, including systems engineering, systems analysis and systems dynamics (Capra 1997).

Post-war era

Cybernetics

While Bertalanffy worked on his general system theory, an interdisciplinary group of mathematician-philosophers, neuroscientists, and engineers, together with a group of humanities scientists, met from 1946 to 1953 in a series of creative New York meetings called the Macy Conferences (Heylighen and Joslyn 2001). Among the eclectic group were the following:

- genius mathematician and philosopher Norbert Weiner;
- quantum theorist, mathematician, game theorist and logician John von Neumann;
- electronic engineer and mathematician Claude Shannon;
- biologist, anthropologist, psychiatrist and epistemologist Gregory Bateson;
- anthropologist Margaret Mead;
- psychologist Kurt Lewin; and
- psychiatrist and physiologist Warren McCulloch (*ibid.*).

From these meetings, the group devised a framework for the emerging field of *cybernetics*,[28] which is concerned with patterns of communication and control in natural and mechanical self-regulating systems. Early cybernetics was built on an odd admixture of earlier Second World War military research and research attempting to express neural mechanisms of mental phenomena in mathematical language (Capra 1997; Heylighen and Joslyn 2001).

According to Capra (1997), *feedback* is the centrepiece of cybernetics which distinguishes it from other types of mechanistic thinking. A basic description of its function, derived from Capra (1997) and Heylighen (2001), is as follows. Feedback is the circular linking of causally connected elements in a system. The circularity of its connections means the first input of the cybernetic system is affected by its last output every time the system cycles, making these systems *non-linear*. When the feedback loop is negative, deviations from equilibrium are suppressed, and the system becomes *self-regulated*. An example of a simple mechanical feedback system is the thermostat. The cyberneticists also turned their minds to feedback cycles in living systems. They believed that self-regulating systems, both living and mechanical, maintained their autonomy and identity (i.e. cohesion) through 'purposeful' action, which is goal directed and apparently intelligent (see Ashby 1964). The cyberneticists also identified feedback as the underlying mechanism of self-regulatory metabolic processes that generate homeostasis in living things. Two types of feedback links were recognised: negative, or self-balancing feedback, and positive, or self-reinforcing feedback, also known as 'run-away' or amplifying feedback. The cyberneticists focused on self-regulating negative feedback systems, those which dampen change and maintain stable states, paying little attention to positive run-away feedback until the 1960s. Cybernetic machines incorporating feedback became a central focus of engineering at this time.

Applied systems thinking

Capra (1997) and Heylighen (2001) also outline how the cyberneticists were interested in communication patterns and building a mathematical theory of communication. They applied probabilistic mathematics used in thermodynamics to measure information transmitted through telegraph and phone lines, and to explain *information entropy*,[29] which led to the formulation of information theory (Shannon and Weaver 1949). They recognised coded messages as patterns of organisation and introduced the term 'bit' as a measurement unit for information. The cyberneticists also devised theory to describe how communication networks (both living and non-living) regulate themselves by generating feedback loops. Communication theory has since made important contributions to electrical engineering and computer science. Cybernetics has also been applied to a variety of other fields, including artificial intelligence, cognitive sciences, neurobiology, molecular evolution, ecological modelling, thermal physics, and human communication in psychology and sociology. In the 1950s, cybernetics also led to a new *cognitive* understanding of brain function (e.g. Broadbent 1958; Putnam 1960). In this view, the brain is understood as a computer-like information-processing machine that uses internal mental models to represent external reality.

Cybernetics and systems thinking also had a strong influence on industrial engineering and management in the 1950s and 1960s, where it was used to solve practical technological problems in systems engineering, systems analysis and systemic management (Capra 1997). Systems analysis had been used by the military in the 1940s for planning and organising operations. It was later adopted by policy makers and technology brokers, and soon morphed into cost-benefit analysis and complex mathematical models for solving organisational problems (*ibid.*). In the 1950s and 1960s Jay Forrester developed the field of systems dynamics, which models the feedback loops, time lags, stocks and flows that affect the overall behaviour of industrial management systems (Forrester 1968). He later applied his ideas to modelling growth in large socio-economic systems in his work *Urban Dynamics* (Forrester 1969) and *World Dynamics* (Forrester 1971). This was later expanded on by Donella Meadows and her collaborators in their controversial book *Limits to Growth* (Meadows *et al.* 1972). In the 1970s, Buzz Holling blended system theories with ecology to form his influential theory of ecological resilience (Holling 1973) and the concept of adaptive management (Holling 1978)—a cybernetic environmental management framework based on two-way feedback between management actions and the state of the resource being managed (see Chapter 1).

Molecular biology

Systems thinking had considerable influence on engineering, organisational management and environmental management through the 1950s, 1960s and 1970s, but had little impact on biology during this time (Capra 1997). In 1944, the physicist Erwin Schrödinger had published his influential book of lectures titled *What is Life?* In the book, Schrödinger applies quantum theory to the process of heredity, postulating that gene molecules are comprised of a small number of atoms combined in multiple ways to produce an almost unlimited number of arrangements in aperiodic crystal-like formation. He further proposes that that these orderly configurations confer genetic molecules with a resistance to becoming disordered by the second law of thermodynamics. Schrödinger's mechanical explanations of life processes launched the field of molecular biology and inspired the likes of Watson and Crick in their search for the hereditary mechanism central to Darwin's theory of evolution. When Watson and Crick finally discovered the structure of DNA in 1953, the emphasis of biological research shifted radically toward molecular biology and genetics.

The discovery of DNA vindicated Weismann's (1893) early theory proposing that germplasm contains hereditary blueprints (genetic determinants) that determine specific phenotypic traits. Denton *et al.* (2013: 42) describe how Weismann's mechanistic theory was in blunt opposition to organicist theories proposing that life's complexity develops through the "*vital epigenetic* activity of the *whole cell-organism* itself".[30] Molecules, as opposed to cells, were now seen as the building blocks of life. The new prevailing view among biologists was that genetic mechanisms not only explained most biological functions, but higher-level phenomena such as behaviour and evolution as well. Any remnant of holistic or systems thinking in biology largely fell away, and favour for mechanism and reductionism resurged as research into the structure and function of DNA progressed. According to Capra (1997: 77), this bias led to a "severe distortion of biological research," the result being 60 years of reductionist molecular approaches that failed to provide adequate explanations for the unique properties of living cells (Denton *et al.* 2013).[31]

Self-organising systems

Ironically, as molecular biologists veered sharply down the reductionist path, cyberneticists were now on a trail that would lead to a revived holistic understanding of both living and non-living systems. In 1943, McCulloch and Pitts (1943) presented a paper depicting neurons as binary switching elements, and a model of the nervous system as a complex network of these elements. Prior to this, cybernetics was primarily concerned with using negative feedback to engineer self-regulating systems. After McCulloch and Pitts's work, however, scientists began building binary network models from which ordered patterns would *spontaneously* emerge out of random behaviour (Capra 1997) in a process called *self-organisation*[32]—a term first introduced then expanded on by Ashby (1947, 1962). In cybernetics, this is referred to as a *step function*, where in passing a critical threshold a system jumps to a completely new way of behaving, defined by a new family of differential equations (Bertalanffy 1969). These models led to the first description of the nervous system as a self-organising system by Ashby (1952). His cybernetic electromagnetic model of brain function demonstrated how systems exhibit adaptive behaviour (learning by trial and error) via step functions. Mathematical descriptions of these self-organising systems, however, were virtually impossible due to their non-linearity and the number and complexity of computations required.

Then, in the 1960s, the development of high-speed computers led to the formulation of non-linear mathematics. This allowed theorists to begin modelling the non-linear aspects of dynamic, self-organising and complex systems that could not previously be modelled. Meteorologist Edward Lorenz (1963) launched the beginning of *chaos theory* with the non-linear statistical models he developed to describe turbulent weather patterns. Chemist and physicist Ilya Prigogine (1967) used non-linear equations to provide the first descriptions of *dissipative systems,* self-organising systems far from equilibrium. From this he devised *non-linear thermodynamics*, the long-awaited mathematical framework that served to underpin Bertalanffy's system theories. Capra (1997) outlines how a number of other scientific breakthroughs followed in short order. Physicist Hermann Haken (1978), for example, developed *synergetics* with his discovery that laser light spontaneously self-organises (emerges) as a result of amplification caused by energising disordered light. Biochemist Manfred Eigen discovered that when energised, enzymatic reactions spontaneously self-organise into networks of catalytic cycles called *hypercycles* (Eigen and Schuster 1977)—the self-organising chemical networks believed to be the molecular precursors to life. In 1972, biologists Humberto Maturana and Francisco Varela (1980), who were also influenced by cybernetics, introduced their theory of *autopoiesis* to describe the circular self-organising nature of living systems. Around the same time, atmospheric chemist James Lovelock and microbiologist Lynn Margulis tackled self-organisation on a planetary level. They used cybernetic principles to devise their Gaia hypothesis (Lovelock and Margulis 1974). The hypothesis identifies a number of complex networks of tightly interlocked feedback loops linking living and non-living systems, which Lovelock and Margulis believed resulted in the earth's self-regulation

Then, in the 1990s, the complexity sciences (see Chapter 4) arose from computer-based explorations into non-linear dynamics and self-organising processes in genetics and biological evolution (e.g. Holland 1987; Kauffman 1993). Heylighen *et al.* (2007) note that despite being a natural extension of cybernetics and systems theory, the chaos and complexity sciences developed largely independently, as a means of finding new ways of predicting and controlling the behaviour of complex systems. As such, the chaos and complexity sciences emerged firmly out of the Newtonian paradigm (*ibid.*) and the tradition of *scientific realism* that had overtaken the logical positivism of the pre-Second World War period.

Scientific realism vs. scientific paradigms

With the Second World War looming, the Vienna Circle dispersed, creating an exodus—mostly to the USA and UK—as many of its members fled or emigrated under fear of mounting Nazism (Uebel 2011). Thus logical positivism spread from its origins in Vienna to have an enormous influence across the Western world, where it remained the dominant philosophy of science until the 1950s, when it was overtaken by *scientific realism*. The logical positivist's purely inductive approach to generating theory had made it impossible to create theory about unobservable entities such as electrons and gravity—from the positivist perspective, unless something could be empirically observed it wasn't real. In addition, their inductive theories were often so broad and flexible they could never be disproved (i.e. they were unfalsifiable; see Chalmers 2002). In reaction to these problems, Karl Popper (1963, 1979), a Vienna-educated philosopher, proposed scientific realism as an alternative. The premise of scientific realism is that science has successfully predicted various phenomena. Hence, the theories used to make these predictions must be true representations of reality (Chalmers 2002). Popper later moderated scientific realism into his *conjectural realism* view of scientific progress (Popper 1979). This view addresses the problems of induction by incorporating the criteria of *falsification* and recognising the fallibility of theory. Popper believed that science works through a process of critical selection and refinement. This involves falsifying old and imperfect theories about the world, and replacing them with new and improved theories that more accurately reflect how the world *really* works.

Despite their differences, positivist and realist philosophies of science are both premised on the materialist assumption that there is an objective independent reality that is being progressively revealed by science (Chalmers 2002). This view was broadly challenged in the 1960s by Thomas Kuhn, with his revolutionary paradigm-based view of scientific progress (see Chapter 7 for detail). Kuhn (1962) showed that science undergoes sporadic revolutionary paradigmatic shifts—whereby crisis causes a predominant theory to be discarded and replaced by another—rather than advancing through gradual progression (whether by falsification or induction) that incrementally reveals the true nature of reality. From this he surmised that scientific reality is largely socially constructed. Despite the strength of Kuhn's work, Popper's conjectural realism and the cumulative notion of scientific progress have remained the dominant view of science since the demise of logical positivism (Chalmers 2002). By demonstrating science's susceptibility to social and cultural influences, Kuhn's ideas did, however, become widely used by *postmodern* thinkers to dispute the objectivity of science.

Postmodern philosophy

Postmodern thought gained popularity in the 1960s, arising as a reaction to the perceived failures of modern industrial society. In the decades that followed it came to have a strong influence on a variety of disciplines, including art, philosophy, economics, politics, and sociology. Postmodernists reject positivistic and scientific assumptions of a single objective reality, and materialist assumptions that 'truth' can be revealed by science. Cilliers (1998: 112) states that modern thinking constitutes an "avoidance of complexity", in its "obsession to find one essential truth". Postmodern philosophy, in contrast, has its roots in the idealist thinking of Kant, who argued we cannot truly know things, as our experiences of them are shaped by our minds (McCormick 2005). As such, postmodern views are commonly premised on the understanding that different groups of people derive different views of reality, as they attempt to draw meaning from their experiences and make sense of their worlds. Because all people exist within unique local circumstances, it is therefore impossible to unify the complexity of their accounts of the world into a single description of reality—a grand narrative or metanarrative (Lyotard 1984). Denunciation of metanarratives[33] forms a basis for postmodern rejection of a single objective reality. Some postmodernists argue that reality is socially constructed because the so-called 'common sense' facts about reality that we take for granted can actually have very different understandings, particularly by those of another culture (see Shotter 1994). As such, we extract our ideas and beliefs about the world through a process of *discourse* with others, using language and communication. Hence, from the postmodern perspective, reality is comprised of a multiplicity of co-existing subjective discourses (Cilliers 1998). Knowledge is understood as the narratives that different groups use to achieve their goals and make sense of their worlds; scientific knowledge, rather than being a unifying source of 'true' knowledge, is viewed as just one such narrative (as per Lyotard 1984).

Quantum spacetime

Meanwhile, in the 80 years since the formulation of quantum mechanics quantum cosmologists have been pursuing their holy grail: *quantum spacetime*, the theory that will unify all physics (Smolin 2002). The notion of whether or not spacetime (i.e. gravity) has an independent existence has been debated since Newton proposed his theories of absolute space and time and was countered by Leibniz's relationist views. The debate largely ended with Einstein's general theory of relativity, which purported that while spacetime/gravity is 'real', it is a dynamical structure that interacts with matter, rather than an inert backdrop in the Newtonian sense. While Einstein's general theory of relativity explains space and time, quantum theory explains everything else in the universe (e.g. elementary particles, atoms, chemistry) (Smolin 2003). However, problems arise when physicists attempt to incorporate relativistic gravity with quantum effects, because in doing so, spacetime appears to dissolve (Wuthrich 2010). As such, physicists have been working on a number of competing theories of quantum spacetime (also referred to as quantum gravity) to resolve this problem, including string theory, co-variant quantum gravity and loop quantum gravity.

Within all these theories, spacetime is viewed not as an ingredient of the universe, or some 'thing' in which the universe is situated, but rather as a higher order phenomena that somehow emerges from a deeper reality that is not spatial or temporal (*ibid.*). This idea, that space and time are not 'real' at a basic level, shatters our conventional understanding of the universe and our idea of physical existence (*ibid.*). As such, some theoretical physicists, such as Lee Smolin (Smolin and Harnad 2008) and Carlo Rovelli (2012), are calling for science to reconsider its long-standing divorce with metaphysics (precipitated by the pre-Second World War logical positivists) in aid of addressing "the deep philosophical and foundational issues in physics" (Smolin and Harnad 2008: 290).

Conclusion

Capra (1997) portrays the history of Western science as a tension and competition between the study of substance (what is it made of?) and the study of form (what is its pattern?). Through this tension Western philosophy, science and mathematics have tortuously co-evolved, with the cyclical emergence and subsidence of opposing ontological theories about the nature of reality: mind versus matter, materialism versus idealism, form (or pattern) versus substance, determinism versus free will, upward versus downward causation, the whole versus the parts, quantity versus quality, and order versus chaos. Over and over these same themes have popped up in different guises and in different arenas. At times these they have mutually reinforced each other, at others broken each other down, in an oscillating dynamic of entwined ideological creation and destruction, played out through history like an epic game of 'whack-a-mole'.

For example, the idea of matter being atomic was first devised by ancient Greek philosophers, shot down by Socratic philosophers, briefly revived by the Epicureans, then, two thousand years later, taken up in the scientific theories of Galileo and Newton. For centuries after, the worldview underlying science was largely materialist and Newtonian, with corresponding philosophies of reductionism, mechanism and modernism (Heylighen *et al.* 2007). Ontologically, this worldview reduces everything to movements of independent material particles governed by deterministic laws (i.e. atomism). Epistemologically, it holds promise of complete, objective and certain knowledge of past and future (*ibid.*). Together, this ontology and epistemology can be understood as the *Newtonian paradigm*, which has underpinned science and Western views for centuries. Its focus on 'being', substance and order led to the creation of the scientific method and revolutionised the world. Indeed, despite modern physics having exploded some key assumptions of the Newtonian paradigm, numerous academic disciplines, including biology, psychology and economics, have uncritically adopted core elements of Newtonian science and methodologies (*ibid.*), many of which are still taught in schools today.

The importance placed on pattern, organising relations and emergence in shaping reality has also emerged and subsided through its history. These ideas were put forward by the Pythagoreans of 500 BC, dismissed by the Socratic philosophers, then revived by German idealist philosophers of the eighteenth century. After receding in the wake of nineteenth century Darwinism and anti-vitalism, interest in pattern resurged once more. General relativity theory and quantum physics revived interest in the holistic philosophies of Kant and Goethe among organicist biologists, holists, process philosophers and quantum physicists of the early twentieth century. The notion of metaphysical wholeness became a centrepiece of Whitehead's process philosophy, Smut's holism, Alexander's emergentism, and Bertalanffy's general system theory. Logical positivism then arose to banish notions of pattern and emergence from much of science and philosophy, just as the Second World War erupted and scattered Europe's intellectual cauldron of great thinkers. Although various fields of applied systems thinking and theories of self-organisation later emerged out of cybernetics, post-war molecular biologists and natural scientists continued to dismiss and neglect holistic notions in favour of reductionism. Meanwhile, as the century progressed, notions of a self-organising emergent reality were revived by postmodern thinking in the social sciences and by the quest for emergent spacetime in the world of physics. They were also independently 'discovered' by chaos and complexity scientists working in the fields of mathematics and evolutionary biology in the latter stages of the century.

In ancient times, Heraclitus purported that all entities come into being through a tension between opposing properties. This view suggests the competing themes that repeatedly emerge, subside and remerge in Western metaphysics may not be in true conflict. Rather, they perhaps represent different aspects of a singular, higher-order phenomenon, just as Newton's absolute space and time were revealed as unified in the fourth dimension by Einstein. In the chapters that follow, I show how mathematical logic can be used to 'prove' this paradoxical idea (Chapter 3), and present concepts from the chaos and complexity sciences to help us understand how this is so (Chapter 4).

Notes

1 In the seventeenth century, as the term 'ontology' came into use, metaphysics became something of a 'catch-all' category; it was later broken down into 'general metaphysics'—the study of being as such, or 'ontology', and 'special metaphysics'—the study of particular types of being, e.g. material, spiritual, etc. (van Inwagen 2007). Since then metaphysics has often been viewed as synonymous with ontology.

2 Which is a variation of the unsourced quote widely attributed to Nobel Prize-winning physicist Richard Feynman: "the philosophy of science is about as useful to scientists as ornithology is to birds".

3 Conversely, Einstein (1949: 684) also noted that epistemology without science is an "empty scheme". He believed that epistemology and science are dependent on each other.
4 Contrary to common belief, the concept of atoms was *invented* by Greek philosophers, rather than discovered through scientific observation of natural phenomena (Goldman 2007).
5 Meaning each planet traced a smaller circular orbit, or epicycle, while following a larger circular orbit around the earth.
6 Calculus was also independently developed by Gottfried Leibniz.
7 Although Newton held the view that many aspects of the world were *not* governed by mathematical laws, and that "God acted upon the world from time to time to preserve it in its required form", holding together an unstable solar system, for example, by preserving the planetary orbits (Hattiangadi 1994: 230).
8 Versus *divine* determinism, whereby events are caused by God (Psillos 2004), which is also seen to exclude free will.
9 Note that many materialists of the time still believed in God; for example, Hobbes came to believe even God was some form of material being (Duncan 2013).
10 Philosophical idealism should not be confused with political or ethical idealism, which is the prioritization of ethics and values over pragmatic concerns.
11 Albrecht (2005: 45) describes the common theme of Hegel's dialectic as "the attempt to present a philosophical account of the distinctive features of organically unified entities, those of identity and continuity despite change and unity, despite difference and diversity".
12 A process or a thing is teleological when it is for the sake of its end.
13 According to Stacey *et al.* (2000), in an attempt to preserve the notion of free will Kant did not apply this formative teleology to humans; instead Kant applies a *rationalist teleology* to humans, whereby they have the freedom to develop their own purpose and therefore create novel outcomes.
14 Whereas Darwin felt the soul was a product of evolution, Wallace (1910) did not. Wallace was similar to the vitalists in his belief that the universe had a purpose, and that supernatural agency was necessary to explain certain aspects of life.
15 Darwin's theory created much debate about teleology, which has carried on to this day. Some have argued that there is no teleology in his theory, as it is based on chance variation (e.g. Dawkins 1986), while others argued that the notion of adaptation is in fact teleological, in that the purpose of an organism is to be adapted to its environment so it can survive (Lennox 1993). Mayr has countered this by stating that that "adaptedness … is a posteriori result rather than an a priori goal-seeking" (Mayr 2004: 58).
16 In an ironic twist, western Europeans largely dismissed the Russian zoologist Pyotr Kropotkin's *Mutual Aid: A factor of Evolution* (1902) as a product of his anarchic views, even though his criticisms of Darwin were founded years before he became an anarchist (Todes 2009).
17 Critics argue that Clements's use of the term 'community' in relation to natural systems is also a case of transference, whereby natural systems are anthropomorphized by applying social theories to them (Foster and Clark 2008).
18 See also Foster and Clark (2008).
19 For a better grasp of this concept, read Edwin Abbott's (1884) *Flatland: A Romance of Many Dimensions*, an amusing introduction to the fourth dimension for the layperson.
20 Thomas (2012) describes the three platforms of Alexander's system. First, spacetime is a 'real' entity (not just a void between things) and identical to matter, which gives rise

material objects through motion. As such, spacetime is a single vast entity of motion. Second, spacetime gives rise to two kinds of characters: categories and qualities (see Chapter 3 to see how this relates to mathematical logic). Third, mind emerges from matter (i.e. spacetime), in a process that can be applied to the universe as a whole, "to produce a hierarchy of existence in which each layer emerges from the next: spacetime leads to matter, matter leads to mind, mind leads to deity" (*ibid.*). While well received at its publication, Alexander's work has since been neglected and is now relatively unknown (*ibid.*).

21 Jan Smuts also established and designed the League of Nations, and was the second prime minister of South Africa. He also has the dubious distinction of having jailed Mohandas Gandhi in 1908.

22 Although he accepted Einstein's four-dimensional spacetime, Whitehead disagreed that spacetime is determined by the distribution of matter.

23 All six of these men were awarded Nobel Prizes for the contributions to quantum theory.

24 A short and simple animated description of this interpretation of the double slit experiment can be viewed at www.youtube.com/watch?v=DfPeprQ7oGc. Note, however, that there are other interpretations of the results, such as that posed by de Broglie and Bohm's pilot wave models.

25 In 1997, two teams separated previously entangled quantum particles, then succeeded in teleporting the quantum state of one particle onto the other particle located elsewhere.

26 While the mathematics of quantum mechanics remained firmly accepted as correct, the debate over what quantum mechanics means in terms of the nature of reality continues. Kumar (2009) describes how the observer-dependence of reality integral to the Copenhagen interpretation creates a problem for cosmology: what happens when the system in question is the whole universe? How does it become real? Without an outside observer the universe can only be a superposition of many possibilities. Attempts to solve this problem have led to the development of several alternative interpretations of quantum mechanics, including the Bohmian, decoherence, and many worlds interpretations. In the many worlds (or multi-verse) interpretation, every possible outcome represented by the wave function exists in an array of parallel universes. This avoids the collapse of the wave function into a single state, and hence the need for an external observer.

27 Now defined as: "a community of organisms and their physical environment interacting as an ecological unit" (Lincoln *et al.* 1998).

28 The term 'cybernetics' is derived from the Greek word for steersman: *kubernetes*. The word was transformed in Latin to *gubernator*, which translates into the English 'governor' (Beer 2002).

29 In the theory, the concept of *information entropy* is introduced as isomorphic to thermodynamic entropy, and as a relative measure of disorder in an information system; conversely, *information redundancy* is used as a measure of relative order.

30 *Epigenetic* in this case refers to factors that do not involve genes.

31 Denton *et al.* (2013) argue that organicist approaches should not be discounted in the twenty-first century. They point to research from the last 20 years showing that epigenetic self-organisation plays an important role in cell formation, and that levels of organisation in cells and organisms do indeed have strange emergent properties that cannot be explained from their constituents.

32 Self-organising systems are characterised by having positive as well as negative feedback loops, and can self-produce (make themselves) and exhibit novel behaviour.

In contrast, self-regulating cybernetic systems are typically externally designed, use negative feedback to dampen change, and cannot on their own reproduce themselves or exhibit new behaviours (Capra 1997).

33 Paradoxically, however, this then means that postmodernity is a metanarrative of 'no metanarrative' (Glover 2006).

References

Abbott, E. A. (1884). *Flatland: A Romance of Many Dimensions*. London: Seeley & Co. Retrieved from https://books.google.ca/books?id=139bAAAAQAAJ&dq=flatland+a+romance+of+many+dimensions&source=gbs_navlinks_s.

Albrecht, G. (2005). Organicism and the organic university. *Concrescence: the Australasian Journal of Process Thought 6*, 43–60.

Alexander, S. (1920). *Space, Time and Deity* (vol. 2). New York: Macmillan.

Allen, G. E. (2005). Mechanism, vitalism and organicism in late nineteenth and twentieth-century biology: the importance of historical context. *Studies in History and Philosophy of Biological and Biomedical Sciences 36*, 261–283.

Ashby, W. R. (1947). Principles of the self-organizing dynamic system. *Journal of General Psychology 37*(2), 125–128.

Ashby, W. R. (1952). *Design for a Brain*. New York: John Wiley.

Ashby, W. R. (1962). Principles of the self-organizing system. In H. v. Foerster and G. W. Zopf (eds), *Principles of Self-Organization: Transactions of the University of Illinois Symposium* (pp. 255–278). London: Pergamon Press.

Ashby, W. R. (1964). *An Introduction to Cybernetics*. London: Methuen.

Aspect, A. (2007). Quantum mechanics: To be or not to be local. *Nature 446*(7138), 866–867.

Barry, J. (1999). *Environment and Social Theory*. New York: Routledge.

Bateson, G. (2000). *Steps to an Ecology of Mind: Collected Essays in Anthropology, Psychiatry, Evolution, and Epistemology*. Chicago, IL: University of Chicago Press.

Bechtel, W. and Richardson, R. (1998). Vitalism. In E. Craig (ed.), *Routledge Encyclopedia of Philosophy*. London: Routledge.

Becker, D. (2004). Kant's moral and political philosophy. In R. C. Solomon and K. M. Higgins (eds), *The Age of German Idealism* (vol. 6, pp. 68–102). New York: Routledge.

Beer, S. (2002). What is Cybernetics? *Kybermetes 31*(2), 209–219.

Berryman, S. (2011). Ancient Atomism. *Stanford Encyclopedia of Philosophy* (Winter 2011 edition). Retrieved from http://plato.stanford.edu/archives/win2011/entries/atomism-ancient.

Bertalanffy, L. V. (1969). *General Systems Theory: Foundations, Development, Applications*. New York: George Braziller.

Bertrand, E. (2010). The balance between determinism and freedom in the philosophy of René Descartes. Paper presented at conference on Chance and Destiny, University of California Santa Barbara, 14–15 May.

Biletzki, A. and Matar, A. (2009). Ludwig Wittgenstein. *Stanford Encyclopedia of Philosophy* (Summer 2011 edition). Retrieved from http://plato.stanford.edu/archives/sum2011/entries/wittgenstein.

Bohm, D. and Hiley, B. J. (1993). *The Undivided Universe: An Ontological Interpretation of Quantum Theory*. London: Routledge.

Bohr, N. (1928). The quantum postulate and the recent development of atomic theory. *Nature 121*, 580–590.

Bohr, N. (1958). *Physics and Human Knowledge*. New York: John Wiley and Sons.

Born, M. (1926). Quantenmechanik der Stoβvorgange. *Z. Phys. 38*, 803.

Bourdeau, M. (2011). Auguste Comte. *Stanford Encyclopedia of Philosophy* (Summer 2011 edition). Retrieved from http://plato.stanford.edu/entries/comte.

Bouwmeester, D., Pan, J.-W., Mattle, K., Eibl, M., Weinfurter, H. and Zeilinger, A. (1997). Experimental quantum teleportation. *Nature 390*(6660), 575–579.

Briggs, J. and Peat, F. D. (1989). *Turbulent Mirror: An Illustrated Guide to Chaos Theory and the Science of Wholeness*. New York: Harper & Row.

Broad, C. D. (1925). *Mind and its Place in Nature*. New York: Harcourt, Brace & Company.

Broadbent, D. E. (1958). Perception and communication. Oxford: Pergamon.

Cannon, W. B. (1932). *The Wisdom of the Body*. New York: W. W. Norton & Company.

Capra, F. (1997). *The Web of Life*. London: Flamingo.

Capra, F. (2010). *The Tao of Physics*. Boston, MA: Shambhala.

Chalmers, A. (2002). *What Is This Thing Called Science?* St Lucia: University of Queensland Press.

Cilliers, P. (1998). *Complexity and Postmodernism: Understanding Complex Systems*. London: Routledge.

Cirne-Lima, C. (1997). *Dialectic for Beginners*. Porto Alegre: Edipucrs.

Clements, F. E. (1905). *Research Methods in Ecology*. Lincoln, NE: University Publishing Company.

Cohen, M. S. (2009). Substances. In G. Anagnostopoulos (ed.), *A Companion to Aristotle* (pp. 197–212). Oxford: Blackwell.

Craig, E. (1998). Metaphysics. *Routledge Encyclopedia of Philososphy Online*. Retrieved from www.rep.routledge.com/article/N095.

D'Espagnat, B. (1979). The quantum theory and reality. *Scientific American* (pp. 158–182) (November).

Darwin, C. (1870). *On the Origin of Species by Means of Natural Selection: Or The Preservation of Favoured Races in the Struggle for Life*. New York: D. Appleton and Company.

Dawkins, R. (1986). *The Blind Watchmaker: Why the Evidence of Evolution Reveals a Universe Without Design*. New York: Norton.

Deichmann, U. and Travis, A. S. (2010). *Darwinism, Philosophy, and Experimental Biology*. New York: Springer.

Denton, M., Kumaramanickavel, G. and Legge, M. (2013). Cells as irreducible wholes: the failure of mechanism and the possibility of an organicist revival. *Biology and Philosophy 28*(1), 31–52.

deVries, W. (2004). Hegel's logic and philosophy of mind. In R. C. Solomon and K. M. Higgins (eds), *The Age of German Idealism* (vol. 6, pp. 216–253). London: Routledge.

Duncan, S. (2013). Thomas Hobbes. *Stanford Encyclopedia of Philosophy* (Spring 2013 edition). Retrieved from http://plato.stanford.edu/archives/spr2013/entries/hobbes.

Eigen, M. and Schuster, P. (1977). The hypercycle: A principle of natural self-organization. *Naturwissenschaften 64*, 541–565.

Einstein, A. (1944) Letter to Robert A. Thornton, 7 December. EA 61-573, Albert Einstein Archives, The Hebrew University of Jerusalem, Israel.

Einstein, A. (1949). Reply to Criticisms. In P. Schilpp (ed.), *Albert Einstein, Philosopher-Scientist*. New York: MJF Books.

Einstein, A., Podolsky, B. and Rosen, N. (1935). Can quantum-mechanical description of physical reality be considered complete? *Physical Review 47*(10), 777.

Elton, C. S. (1927). *Animal Ecology*. London: Sidgwick & Jackson.

Falcon, A. (2012). Aristotle on Causality. *Stanford Encyclopedia of Philosophy* (Winter

2012 edition). Retrieved from http://plato.stanford.edu/cgi-bin/encyclopedia/archinfo.cgi?entry=aristotle-causality.

Ferraro, R. (2007). *Einstein's Space-Time: An Introduction to Special and General Relativity*. New York: Springer.

Forrester, J. W. (1968). *Principles of Systems*. Waltham, MA: Pegasus Communications.

Forrester, J. W. (1969). *Urban Dynamics*. Waltham, MA: Pegasus Communications.

Forrester, J. W. (1971). *World Dynamics*. Waltham, MA: Pegasus Communications.

Foster, J. F. and Clark, B. (2008). The sociology of ecology: ecological organicism versus ecosystem ecology in the social construction of ecological science, 1926–1935. *Organization and Environment 21*(3), 311–352.

Furley, D. (1999). Aristotle the philosopher of nature. In D. Furley (ed.), *Routledge History of Philosophy: from Aristotle to Augustine* (vol. 2). London: Routledge.

Gilbert, S. F. and Sarkar, S. (2000). Embracing complexity: organicism for the 21st century. *Developmental Dynamics 219*, 1–9.

Gleason, H. A. (1926). The individualistic concept of the plant association. *Bulletin of the Torrey Botanical Club 53*(1), 7–26.

Glover, L. (2006). *Postmodern Climate Change*. New York: Routledge.

Goldman, S. (2007). *Great Scientific Ideas that Change the World: Part 1* (vol. 1). Chantilly, VA: The Teaching Company.

Graham, D. (2011). Heraclitus. *The Stanford Encyclopedia of Philosophy* (Summer 2011 edition). Retrieved from http://plato.stanford.edu/archives/sum2011/entries/heraclitus.

Graham, D. (2012). Heraclitus as Process Philosopher. *Philosophy Study 2*(1), 1–8.

Gribbin, J. (1984). *In Search of Schrödinger's Cat: Quantam Physics and Reality*. New York: Bantam Books.

Griffin, D. R. (1988). Of minds and molecules: Postmodern medicine in a psychosomatic universe. In D. R. Griffin (ed.), *The Reenchantment of Science: Postmodern Proposals* (pp. 141–163). Albany, NY: State of New York Press.

Gustavsson, K. (2010). Charles Dunbar Broad. *Stanford Encyclopedia of Philosophy* (Fall 2010 edition). Retrieved from http://plato.stanford.edu/archives/fall2010/entries/broad.

Haken, H. (1978). *Synergetics: An Introduction* New York: Springer-Verlag.

Harrington, A. (1996). *Reenchanted Science: Holism in German Culture from Wilhelm II to Hitler*. Princeton, NJ: Princeton University Press.

Hatfield, G. (2008). Rene Descartes. *Stanford Encyclopedia of Philosophy,* (Summer 2011 edition). Retrieved from http://plato.stanford.edu/archives/sum2011/entries/descartes.

Hattiangadi, J. (1994). Philosophy of biology in the nineteenth century. In G. H. Parkinson and S. G. Shanker (eds), *The Nineteenth Century* (pp. 228–247). London: Routledge.

Heisenberg, W. (1927). Zeit fur Phys. *1936 Funf Wiener Vortage 43*.

Heylighen, F. and Joslyn, C. (2001). Cybernetics and Second Order Cybernetics. In R. A. Meyers (ed.), *Encyclopedia of Physical Science and Technology* (pp. 155–170). New York: Academic Press.

Heylighen, F., Cilliers, P. and Gershenson, C. (2007). Complexity and philosophy. In J. Bogg and R. Geyer (eds), *Complexity, Science and Society* (pp. 117–134). New York: Radcliffe Publishing.

Holland, J. H. (1987). Complex adaptive systems: A primer. *Santa Fe Institute Bulletin 2*(1), 8–10.

Holling, C. S. (1973). Resilience and stability of ecological systems. *Annual Review of Ecology and Systematics 1973*, 1–23.

Holling, C. S. (1978). *Adaptive Assessment and Management*. Chichester: Wiley.

Huggett, N. and Hoefer, C. (2009). Absolute and relational theories of space and motion. *Stanford Encyclopedia of Philosophy* (Fall 2009 edition). Retrieved from http://plato.stanford.edu/archives/fall2009/entries/spacetime-theories.

Jeans, J. (1930). *Mysterious Universe*. New York: Cambridge University Press.

Jeans, J. (1934). The New World-picture of Modern Physics. An address to the British Association for the Advancement of Science, Capital Cinema, Aberdeen.

Kauffman, S. (1993). *The Origins of Order: Self-Organization and Selection in Evolution*. New York: Oxford University Press.

Keller, D. R. and Golley, F. B. (2000). *The Philosophy of Ecology: From Science to Synthesis*. Athens, GA: University of Georgia Press.

Kirk, G. S. and Raven, J. E. (1983). *The Presocratic Philosophers: A Critical History with a Selection of Texts*. Cambridge: Cambridge University Press.

Kuhn, T. S. (1962). *The Structure of Scientific Revolutions*. Chicago, IL: University of Chicago Press.

Kumar, M. (2009). *Quantum: Einstein, Bohr and the Great Debate About the Nature of Reality*. London: Icon Books.

Laplace, P.-S. (1951). *A Philosophical Essay on Probabilities* (trans. F. W. Truscott and F. L. Emory). New York: Dover Publications.

Lennox, J. G. (1993). Darwin was a teleologist. *Biology and Philosophy 8*, 409–421.

Levins, R. and Lewontin, R. (1985). *The Dialectical Biologist*. Cambridge MA: Harvard University Press.

Lincoln, R. J., Boxshall, G. A. and Clark, P. F. (1998). *A Dictionary of Ecology, Evolution and Systematics*. Cambridge: Cambridge University Press.

Lorenz, E. N. (1963). Deterministic nonperiodic flow. *Journal of the Atmospheric Sciences 20*(2), 130–141.

Lorimer, D. (2013). Metaphysical foundations of modern science. Retrieved 12 August 2013 from www.scimednet.org/metaphysical-foundations-of-modern-science.

Lovelock, J. E. and Margulis, L. (1974). Atmospheric homeostasis by and for the biosphere: the Gaia hypothesis. *Tellus 26*(1–2), 2–10.

Lyotard, J. F. (1984). *The Post-Modern Condition: A Report on Knowledge*. Minnesota, MN: University of Minn. Press.

Malone-France, D. (2007). *Deep Empiricism: Kant, Whitehead, And the Necessity of Philosophical Theism*. Lanham, MD: Lexington Books.

Mastin, L. (2008). Monism. *The Basics of Philosophy*. Retrieved August 20 2012 from www.philosophybasics.com/branch_monism.html.

Maturana, H. and Varela, F. J. (1980). *Autopoiesis and Cognition: The Realization of the Living*. Dordrecht: Reidel.

Mayr, E. (1988). *Toward a New Philosophy of Biology: Observations of an Evolutionist*. Cambridge, MA: Belknap.

Mayr, E. (1999). Darwin's influence on modern thought. Paper presented at the Lecture delivered in Stockholm on receiving the Craafoord Prize, Royal Swedish Academy of Science, 23 September. Retrieved from www.biologie.uni-hamburg.de/b-online/e36_2/darwin_influence.htm.

Mayr, E. (2004). *What Makes Biology Unique? Considerations on the Autonomy of a Scientific Discipline*. Cambridge: Cambridge University Press.

McCormick, M. (2005). Immanuel Kant: metaphysics. *Internet Encyclopedia of Philosophy*. Retrieved from www.iep.utm.edu/kantmeta.

McCulloch, W. S. and Pitts, W. (1943). A logical calculus of the ideas immanent in nervous activity. *The Bulletin of Mathematical Biophysics 5*(4), 115–133.

McDonough, R. (2010). Plato's Organicism. *Internet Encyclopedia of Philosophy*. Retrieved from www.iep.utm.edu/platoorg.

Mcintosh, D. (2012). Plato: a theory of forms. *Philosophy Now* (90). Retrieved from http://philosophynow.org/issues/90/Plato_A_Theory_of_Forms

McQuillan, C. (2012). German idealism. *Internet Encyclopedia of Philosophy*. Retrieved from www.iep.utm.edu/germidea.

Meadows, D. H., Meadows, D. L., Landers, J. and Behrens III, W. W. (1972). *The Limits to Growth*. New York: Universe Books.

Moller, P. H. (2006). Holism and Evolution: Jan Smuts and his permanent mission. Dedication to 80th anniversary of the first publishing of Holism and Evolution at London, 1926. University of South Bohemia.

Morgan, C. L. (1923). *Emergent Evolution*. London: Williams and Norgate.

Musser, G. (2013). The wholeness of quantum reality: an interview with physicist Basil Hiley. *Critical Opalescence* (vol. 2015). Retrieved from http://blogs.scientificamerican.com/critical-opalescence/the-wholeness-of-quantum-reality-an-interview-with-physicist-basil-hiley.

Nixon, G. M. (2010). Whitehead and the elusive present: Process philosophy's creative core. *Journal of Consciousness Exploration and Research 1*(5), 625–639.

Overduin, J. (2007). Einstein's spacetime. *Gravity Probe B: Testing Einstein's Universe*. Retrieved 2 April 2015 from https://einstein.stanford.edu/SPACETIME/spacetime2.html.

Pajares, G. (2012). Aristotle as an idealist. *JPAIR Multidisciplinary Journal 7*, 133–143.

Phillips, J. (1931). The biotic community. *Journal of Ecology 19*, 1–24.

Popper, K. (1963). *Conjectures and Refutations: The Growth of Scientific Knowledge*. London: Routledge.

Popper, K. (1979). *Objective Knowledge: An Evolutionary Approach*. Oxford: Oxford University Press.

Prigogine, I. (1967). Dissipative structures in chemical systems. In S. Claesson (ed.), *Fast Reactions and Primary Processes in Chemical Kinetics*. New York: Interscience.

Prigogine, I. and Stengers, I. (1984). *Order Out of Chaos: Man's New Dialogue with Nature*. New York: Bantam Books.

Prinz, J. (ed.) (2006). *Concise Encyclopedia of Philosophy of Language and Linguistics*. Oxford: Elsevier.

Psillos, S. (2004). Causation. In M. C. Horowitz (ed.), *New Dictionary of the History of Ideas* (pp. 272–280). New York: Charles Scribner's Sons.

Putnam, H. (ed.). (1960). *Minds and Machines*. New York: New York University Press.

Rabin, S. (2010). Nicolaus Copernicus. *Stanford Encyclopedia of Philosophy* (Fall 2010 edition). Retrieved from http://plato.stanford.edu/archives/fall2010/entries/copernicus.

Rescher, N. (2008). Process Philosophy. *Stanford Encyclopedia of Philosophy* (Summer 2012 edition). Retrieved from http://plato.stanford.edu/archives/sum2012/entries/process-philosophy.

Rickles, D. (2008). Quantum Gravity: A Primer for Philosophers. In D. Rickles (ed.), *The Ashgate Companion to Contemporary Philosophy of Physics*. Farnham: Ashgate Publishing.

Ridling, Z. (2001). *Philosophy Then and Now: A Look Back at 26 Centuries of Ideas That Have Shaped our Thinking*. Access Foundation.

Ritchie, J. and Lewis, J. (2003). *Qualitative Research Practice: A Guide for Social Science Students and Researchers*. London: Sage Publications.

Robinson, H. (2011). Dualism. *Stanford Encyclopedia of Philosophy* (Winter 2012 edition). Retrieved from http://plato.stanford.edu/archives/win2012/entries/dualism.

Rogers, G. (2004). Science and British Philosophy: Boyle and Newton. In S. Brown (ed.), *British Philosophy and the Age of Englightenment* (pp. 35 55). London: Routledge.

Rovelli, C. (2012). Science is not about certainty: A philosophy of physics. *Edge*. Retrieved 15 September 2012 from https://edge.org/conversation/a-philosophy-of-physics.

Ruse, M. (2013). Philosophy of biology. *Encyclopedia Brittannica Online*. Retrieved 4 February from www.britannica.com/EBchecked/topic/681551/biology-philosophy-of/283537/History.

Rynasiewicz, R. (2004). Newton's Views on Space, Time and Motion. *Stanford Encyclopedia of Philosophy* (Winter 2012 edition). Retrieved from http://plato.stanford.edu/archives/win2012/entries/newton-stm.

Schrödinger, E. (1944). *What is Life? The Physical Aspect of the Living Cell*. Cambridge: Cambridge University Press.

Seibt, J. (2013). Process philosophy. *The Stanford Encyclopedia of Philosophy* (Fall 2013 edition). Retrieved from http://plato.stanford.edu/archives/fall2013/entries/process-philosophy.

Shanker, S. G. (1996a). Descartes' legacy: the mechanist/vitalist debate. In S. G. Shanker (ed.), *Routledge History of Philosophy* (vol. IX, pp. 315–375). London: Routledge.

Shanker, S. G. (ed.). (1996b). *Philosophy of Science, Logic and Mathematics in the Twentieth Century* (vol. IX). London: Routledge.

Shannon, C. and Weaver, W. (1949). *The Mathematical Theory of Communication*. Urbana, IL: University of Illinois Press.

Sherburne, D. W. (1981). *A Key to Whitehead's Process and Reality*. Chicago, IL: University of Chicago Press.

Shields, C. (2011). Artistotle. *Stanford Encyclopedia of Philosophy* (Summer 2012 edition). Retrieved from http://plato.stanford.edu/archives/sum2012/entries/aristotle.

Shotter, J. (1994). *Conversational Realities: Constructing Life through Language*. London: Sage.

Silverman, A. (2008). Plato's Middle Period Metaphysics and Epistemology. *Stanford Encyclopedia of Philosophy* (Summer 2012 edition). Retrieved from http://plato.stanford.edu/archives/sum2012/entries/plato-metaphysics.

Skrbina, D. (2007). Panpsychism. *Internet Encyclopedia of Philosophy*. Retrieved from www.iep.utm.edu/panpsych.

Smith, G. (2007). Isaac Newton. *Stanford Encyclopedia of Philosophy* (Fall 2008 edition). Retrieved from http://plato.stanford.edu/archives/fall2008/entries/newton.

Smith, K. (2010). Descartes' life and works. *Stanford Encyclopedia of Philosophy* (Fall 2012 edition). Retrieved from http://plato.stanford.edu/archives/fall2012/entries/descartes-works.

Smolin, L. (2002). *Three Roads To Quantum Gravity*. New York: Basic Books.

Smolin, L. (2003). Loop quantum gravity: Lee Smolin. *Edge*. Retrieved from http://edge.org/conversation/loop-quantum-gravity-lee-smolin.

Smolin, L. and Harnad, J. (2008). The trouble with physics: The rise of string theory, the fall of a science, and what comes next. *The Mathematical Intelligencer 30*(3), 66–69.

Smuts, J. C. (1927). *Holism and Evolution*. Toronto: McMillan Company.

Solomon, R. C. and Higgins, K. M. (eds). (2004). *The Age of German Idealism* (vol. 6). New York: Routledge.

Sorensen, R. (2012). Nothingness. *The Stanford Encyclopedia of Philosophy* (Winter 2012 edition). Retrieved from http://plato.stanford.edu/archives/win2012/entries/nothingness.

Stacey, R., Griffin, D. and Shaw, P. (2000). *Complexity and Management: Fad or Radical Challenge*. New York: Routledge.

Stapp, H. P. (1971). S-matrix interpretation of quantum theory. *Physical Review D 3*(6), 1303.
Stern, R. (1990). *Hegel, Kant and the Structure of the Object*. New York: Routledge.
Stokes, P. (2012). *Philosophy 100 Essential Thinkers: The Ideas that have Shaped our World*. London: Arturus Publishing.
Tamayao, M. (2012). Scholastic Philosophy. *Weblog for Mike's Lectures and Papers*. Retrieved 15 January 2013 from http://tamayaosbc.wordpress.com/2012/01/29/scholastic-philosophy.
Tansley, A. G. (1935). The use and abuse of vegetational concepts and terms. *Ecology 16*(3), 284–307.
Thomas, E. (2012). Samuel Alexander (1859–1938). *Internet Encyclopedia of Philosophy*. Retrieved from www.iep.utm.edu/alex-sam.
Todes, D. P. (2009). Global Darwin: Contempt for competition. *Nature 462*(5), 36–37.
Uebel, T. (2011). Vienna Circle. *Stanford Encyclopedia of Philosophy* (Summer 2012 edition). Retrieved from http://plato.stanford.edu/cgi-bin/encyclopedia/archinfo.cgi?entry=vienna-circle.
Van Inwagen, P. (2007). Metaphysics. *Stanford Encyclopedia of Philosophy* (Winter 2012 edition). Retrieved from http://plato.stanford.edu/archives/win2012/entries/metaphysics.
Van Melsen, A. (2015). Atomism. *Encyclopaedia Britannica*. Retrieved from www.britannica.com/EBchecked/topic/41810/atomism/68632/Atoms-in-external-aggregation-versus-in-internal-relationship.
Wallace, A. R. (1871). *Contributions to the Theory of Natural Selection: A Series of Essays*. New York: McMillan and Co.
Wallace, A. R. (1910). New thoughts on evolution. *The Daily Chronicle* (3–4 November). Retrieved from http://people.wku.edu/charles.smith/wallace/S746.htm.
Weber, M. and Esfeld, M. (2003). Holism in the sciences. In *Unity of Knowledge in Transdisciplinary Research for Sustainability: Encyclopedia of Life Support Systems*. Oxford: EOLLS Publishers.
Weismann, A. (1893). *Germ-plasm: A Theory of Heredity* (vol. 2). New York: C. Scribner's Sons.
Wheeler, A. (1973). From relativity to mutability. In J. Mehra (ed.), *A Physicist's Conception of Nature* (pp. 202–247). Boston, MA: D. Reidel Publishing Company.
Whitehead, A. N. and Russell, B. (1913). *Principia Mathematica*. Cambridge: Cambridge University Press.
Wittgenstein, L. (1922). *Tractatus Logico-philosophicus* (trans. C. K. Ogden). London: Routledge & Kegan Paul.
Wudka, J. (2006). *Space-Time, Relativity, and Cosmology*. Cambridge: Cambridge University Press.
Wuthrich, C. (2010). Emergent spacetime in quantum theories of gravity. Retrieved 20 September 2012 from http://philosophyfaculty.ucsd.edu/faculty/wuthrich/research/emergentspacetime.html.
York, J. (2013). Pythagorean natural philosophy. Retrieved 4 February 2013 from www.jamieyorkpress.com/wp-content/uploads/2012/04/Pythagorean-math-philosophy.pdf.
Zeyl, D. (2014). Plato's Timaeus. *The Stanford Encyclopedia of Philosophy* (Spring 2014 edition). Retrieved from http://plato.stanford.edu/archives/spr2014/entries/plato-timaeus.

3 Unravelling paradox

Mathematical logic and orders of change

Collections: wholes and not wholes brought together, pulled apart sung in unison, sung in conflict from all things one and from one all things.

(Heraclitus[1])

The logic of paradox

In ancient times Heraclitus posited that that all entities come into being through a tension between opposing properties, that opposites, though different, are nonetheless interconnected. The discussion in Chapter 2 illustrates how this metaphysical thread has woven in and out of Western science and philosophy throughout history. Nonetheless, for most contemporary Western minds, Heraclitus's ideas are paradoxical and counterintuitive.

Slater (2005) describes paradox as "a puzzling conclusion we seem to be driven towards by our reasoning, but which is highly counterintuitive, nevertheless". Most paradox involves contradiction. Northrop (2014) describes it is anything which appears to be true, but which is actually false; or which appears to be false and is actually true. To appreciate the maddening mental effect of a simple circular paradox, write this on a slip of paper: "the statement on the other side is TRUE". On the reverse side of the paper write "the statement on the other side is FALSE". Then experience the circular confusion that ensues as you try to follow the logic of the statements while reading first one side of the paper, then the other.

When a paradox such as this is revealed, it signals a hidden error in how a problem or a statement has been formulated. The above paradox, for example, reveals the error of an argument structured so its premises are in need of validation as much as its conclusion. This relates back to wicked problems and Watzlawick *et al.*'s (2011) principles of problem formulation, which state that intractable problems are typically created and perpetuated by the way they have been framed. Dell (1981) notes that paradox is inherently epistemological: it emerges out of the premises and beliefs of an observer—there are no phenomena that are paradoxical in themselves. Thus, paradoxical situations are flashing signs, warning that a concealed invalid argument is at play. Which brings us back to Heraclitus. Despite the validation general relativity and quantum theory garner for Heraclitus's ideas about the nature of reality, his metaphysics are intuitively paradoxical to us, nonetheless. This signifies that the way we've framed our understanding of reality may be flawed, that there is a logical error buried somewhere in our assumptions of how the universe works.

Because paradoxical errors are hidden and counterintuitive, revealing and resolving them requires mental gymnastics best performed by highly trained minds. For centuries, Taoist masters have used paradox to "expose the inconsistencies arising from verbal communication and to show its limits" (Capra 2010: 43). Zen masters have likewise used it to hone the minds of their students and transcend the self-referential limits of everyday language and rationality (Fitzpatrick 2005; Tarrant 2005). In the West, however, the master unravellers of paradox have been mathematical logicians. Mathematical logic is concerned with formal logic and proof systems, and increasing precision in reasoning. In the 1800s, mathematicians began recognising a number of logical inconsistencies and gaps in their trade. They started applying formal logic to prove the consistency of mathematical foundations and to develop axiomatic frameworks for geometry, arithmetic and analysis (Irvine 2010). In doing so, however, they revealed a number of paradoxical 'cracks' in the basics of math and logic. The early twentieth century saw the era's greatest mathematical minds—Henri Poincaré, Bertrand Russell, Alfred North Whitehead and Kurt Gödel—bear down on resolving these paradoxes, to correct errors in how the rules and premises of mathematics and logic had been formulated (Cantini 2012).

The mathematical logic developed in this process has since been used to help solve problems in other fields, including philosophy, cybernetics, computer sciences, linguistics, neuroscience, psychology and the complexity sciences. In this short chapter I draw on the works of Watzlawick *et al.* (2011) and Bateson (2000) to introduce three logic theories that can be used to untangle paradoxical situations and seemingly intractable problems: group theory, the theory of logical types, and Gödel's incompleteness theorem.

Group theory

One evening in May of 1832, the young mathematician Evariste Galois, while under threat of death in a sunrise duel, frantically penned group theory. He was killed the next morning, just 20 years old, leaving behind a mathematical legacy that helped lay the foundations of abstract algebra and relativity and quantum theory (Watzlawick *et al.* 2011). Galois's group theory is a mathematical framework for thinking about change among members within a class or group, and explains how persistence is interrelated with change. Watzlawick *et al.* note that groupings are the most basic element of human perception, in that they give structure to what would otherwise seem a chaotic environment. The authors also provide detail of how the properties outlined by Galois's group theory demonstrate the "peculiar interdependence between persistence and change" in groups (*ibid.*: 7). Galois's group theory, coarsely interpreted, postulates that a *group* has four properties:

1. It is composed of members that all share one common characteristic; members can be numbers, objects, concepts, etc., as long as the outcome of combining any members of the group is itself a member of the group.
2. Its members can be combined in any sequence, and the outcome will be the same.
3. A group contains an *identity member*, which, when combined with any other member, gives that other member; i.e. the other member maintains its identity.
4. Every member in the group has an opposite, and when the two are combined, they give the identity member.

(Watzlawick *et al.* 2011)

For example, in relation to the first property, Watzlawick *et al.* use the example of members of a group consisting of the numbers 1–12, indicating the hours on a clock. If 12:00pm is combined with six hours, the result is 6:00pm, or if combined with twelve hours the result is 12:00am; both results, 6:00pm and 12:00am, are themselves group members. Thus, when members of the group are combined, the result is a change from one *internal* state of the group to another; however, the overall group itself is unchanged. Ordering things into groups establishes *invariance*, such that any combination of members results in yet another group member. The result will never be a member that does not fit inside the group; the result is always a member that shares the common group characteristic.

To demonstrate the second property, Watzlawick *et al.* use a group containing a number of position changes (or moves) on a surface, each having a particular length and direction. No matter how these moves are combined, one will always end up in the same final position. The simplest case would be one where the group consists of four moves, each of the same length, in each of the four cardinal directions, north, south, east and west. Regardless how these moves are ordered, one always finishes in the same position—in this case where one started (try it on a piece of paper; then try it with a group consisting of moves of different lengths and directions). Thus, there is "changeability in process, but invariance in outcome" (*ibid.*: 7). No matter how the group is reorganised, the final result is always the same.

Watzlawick *et al.* illustrate the third property as follows. When the combination rule for a group is additive, the group's *identity member* is 0. For example, 5 + 0 = 5. When the given member (5) is combined with the identity member (0), the result is the given member (5). If the group's combination rule is multiplication, however, then the identity member is 1, for example 5 × 1 = 5. Another example is a group which has sound as its common characteristic (i.e. invariant property). The identity member for this group would be *silence*—when a sound is combined with silence, the outcome is sound. If the group's common characteristic is motion, then the identity member would be stillness. Although the notion of an identity member seems pointless, it is a special case of group invariance that serves an important 'null function' in maintaining group stability. The point here is that however a group's identity member acts or combines with other group members, it makes no difference to the final outcome, the relevance of which is demonstrated in the next example.

Finally, with respect to the fourth property, Watzlawick *et al.* again use an example of a group where the combination rule is additive. If you take group member 5, combine it with its opposite –5, the result is 0, which is the group's identity number: 5 + –5 = 0. Thus, although combining a group member with its opposite constitutes a large change, that change results in the group's identity member, which has no effect on the final outcome.

These examples demonstrate how marked changes taking place *within* the group do not cause any change in the *overall* group itself. When a system runs through all its possible internal states without effecting an overall systemic change, it has become caught in a *game without end* (see Watzlawick *et al.* 1967). The system cannot generate from *within* the conditions needed to change its own rules. *Deadlocks* are another result of trying to effect change from within a system. The theory also relates to circular negative feedback systems that maintain system stability in cybernetic and homeostatic processes (Watzlawick *et al.* 2011).

Watzlawick *et al.* evoke Galois's group theory to explain the paradoxical relationship between persistence and change; why, despite their apparent opposing natures, they must occur together. As such, Galois's group theory helps resolve the paradox of Heraclitus's unity of opposites, and the Pythagorean notion of reality coming forth through the interplay of 'being' and 'not being', a concept later built on by Hegel in the Romantic era. All perception is necessarily relative, a matter of contrast and comparison—nothing can exist without its opposite. If everything in the universe is blue, it is not possible to have a concept of 'blueness' (*ibid.*). Likewise, if there is no sleeping it is not possible to have a concept of being awake; if there is no noise, there can be no silence. This notion relates to the Zen koan that asks, "What is the sound of one hand clapping?" Watzlawick *et al.* (*ibid.*) posit that although many theories of change and persistence have been formulated in Western culture, in the last centuries most of these have been about either change or persistence, not the interrelationship between the two.

Theory of logical types

Between 1910 and 1927, Alfred Whitehead and his student Bertrand Russell published the seminal *Principia Mathematica* (1913), an attempt to build a logically consistent set theory on which to found all mathematics. Their book is possibly the most important contribution to logic and philosophy since Aristotle, and one of the twentieth century's greatest scientific documents (Irvine 2010). Not only did *Principia Mathematica* demonstrate the deductive power of logic, it also re-established the links between logic, metaphysics and epistemology, and influenced a wide range of fields from math and philosophy to computer science, linguistics and psychology (*ibid.*). Much of the book is dedicated to resolving *Russell's Paradox*, a contradiction Russell discovered in set theory whilst working on the book (see Clement 2005).

Paradoxes have long baffled Western thinkers. Watzlawick *et al.* (2011) use a classic example related to Russell's paradox: the ancient Epimenides paradox whereby a Cretan gives the warning that "all Cretans are liars". If the statement is true, it is false; if it is false, it is true. In self-referentially asserting its own falseness, the sentence creates a paradox. Such statements are not provable and create a circular confusion in the minds of those who encounter them. This powerful effect is one of the bases of humour—comedians create contradictions in the minds of an audience, which in turn generates laughter (Cave 2005). Groucho Marx, for example, used a version of the Cretan paradox when he joked: "Please accept my resignation. I don't want to belong to any club that will accept people like me as a member" (Marx 1959: 321).

In *Principia Mathematica*, Russell and Whitehead introduce their theory of logical types, which they devised to rule out self-reference in logic and prevent emergence of paradox (specifically Russell's Paradox). The theory starts with the idea of a collection consisting of 'things' united by a common characteristic (i.e. a group). Whitehead and Russell use the term 'member', as did Galois, to refer to the individual 'things' making up a collection. However they use the term 'class', rather than 'group', to refer to the total collection of members. According to the theory, "whatever involves all of a collection must not be one of the collection", as they are of different *logical types*, or levels (Watzlawick *et al.* 2011: 8).

Watzlawick *et al.* illustrate the notion of logical types/levels using the concept of 'humankind'. Humankind is the class of all human individuals, but is not itself an individual human; i.e. humans and humankind are different logical types. As such, any attempt to deal with one in terms of the other—confusing logical types—will lead to "nonsense and confusion" (*ibid.*). The authors use the so-called Robinson Crusoe economic model as another example of confusing logical levels. The model assumes that the economic behaviour of a large city can be explained by taking the economic behaviour of one of its inhabitants and multiplying it by the city's population. The critical distinction here is that a city of four million is not just *quantitatively* different from an individual, it is *qualitatively* different (i.e. a class is qualitatively different from its members). Errors in logical typing normally occur either because a particular property has been incorrectly ascribed to a class instead of a member (or vice versa), or by treating class and member as if they were on the same level of abstraction (*ibid.*). Such errors result in paradox and *vicious circles.*[2] The theory of logical types resolves the 'all Cretans are liars' paradox by asserting that the Cretan who calls other Cretans liars cannot be included in the class of other Cretans in this context. By making a statement concerning the whole of a group of people, he cannot be *of* that group of people; therefore statements he makes about the group cannot be applied to him (Grof 1981).

Watzlawick *et al.* (2011) suggest that we are universally faced with hierarchies of logical levels created by classes and their members, and that our frequent confusion of logical levels leads to ever-present "puzzling consequences" (*ibid.*: 9). A class cannot be understood in the language of its members; rather a new *meta-language* is required. The authors note that the phenomenon of change is subject to this principle, be it in physics or in human behaviour. This can be illustrated by elaborating on the example of position, velocity and acceleration used by both Bateson (2000) and Watzlawick *et al.* (2011). Begin by thinking about the notion of *position*. Position is described using the framework of space and understood mathematically as a point corresponding to coordinates on an *x*, *y*, *z* axis. Then introduce the idea of a *change in position*. In order to talk about or understand a change in position, one must step outside the theoretical framework of *position*. This is because the idea of a changing position cannot be generated using the language or computational framework (e.g. a coordinate system) used to describe position. Rather the dimension of *time* must be introduced and with it a whole new language (motion, velocity) and computational framework is needed (e.g. algebra).[3] In other words, shifting one's thought from the notion of position to that of a change in position requires a *jump* to higher logical level, or what cyberneticists refer to as a step function. A change in motion, being a change in a change of position, is a *meta-change*, and introduces more new language (acceleration, deceleration) and computations (e.g. calculus). A change in acceleration is a meta-meta-change, and again, requires a new language (jerk) and computational framework for its description (one which created enormous theoretical challenges for early space scientists).[4] Attempting to deal with change in position (motion) within the framework of position, or a change in motion (acceleration/deceleration) within the framework of motion, violates the axiom of the theory of logical types and leads to paradoxical confusion.

Thus, a new language and computational framework are required for describing phenomena at each logical level. Explorations into quantum physics were only opened up by the invention of a new mathematical language incorporating the concept of *operators*, as classical formulations for explaining macroscopic phenomena cannot be applied at the quantum level. The hierarchy of logical types can also be thought of in relation to Einstein's notion of four-dimensional spacetime. The intersection of latitude and longitude, two independent coordinates, specify a unique *position*, or point, on the earth's two-dimensional surface. Add depth, or height, and *objects* can be specified in three-dimensional space. Then add time, and an *event* is now specified in four-dimensional spacetime. Each shift to a higher dimension requires the introduction of a qualitatively new concept (first space, then time), and therefore a jump to a higher logical level.

If we make an error in logical typing when attempting to explain phenomena, paradoxes are generated. For example, by treating space and time as absolute and unrelated (see Figure 2.1 in Chapter 2), Newton's universal theory of gravitation generated the paradox of instantaneous gravitational force, requiring information to travel faster than the speed of light in order to allow masses to respond to changes in gravitational field. This paradox was finally resolved when Einstein unified space and time into a curved four-dimensional spacetime, thereby assigning it a higher logical order. The paradoxical nature of Whitehead and Russell's theory is also more simply exemplified by the standard metre in Paris: it is the one item in the world that cannot be measured using the metric system, because it is the basis of the entire metric system (Watzlawick *et al.* 2011).

Watzlawick *et al.* (2011) also provide language-based examples that illustrate this theory. There are many different things expressed in a language, but in order to talk about the language itself, linguists have had to invent a meta-language, which also needs a meta-meta-language to discuss its structure. Another example is *methodology*, the philosophical study of the methods used in different disciplines, which is of a higher logical type than *method*, and therefore a meta-method. Watzlawick *et al.* note that our language often creates confusion in failing to discern between logical types, and thereby assigning the same names to things that are of different logical levels. Bateson (2000) uses the example of 'wave', which describes both a class of particle movements and a moving ridge of water, which are qualitatively different things of different logical levels. We may say a wave moves, but what we are actually referring to is a meta-movement, a movement of a class of movements[5] (Figure 3.1) to which different rules and premises apply. For example, friction works to slow the movement of water particles, but does not slow the meta-movement of a ridge of water.

Watzlawick *et al.* (2011) sum up by drawing two key conclusions from the theory of logical types: 1) to prevent paradox and confusion logical levels must be strictly separated; and 2) moving between logical levels involves a discontinuous 'jump' of some sort, a transformation. This jump is of practical importance as it provides a route *out of the system*, the relevance of which is discussed in the chapters that follow.

Gödel's incompleteness theorem

In 1931, Kurt Gödel, a member of the Vienna Circle and widely regarded as the most important (and perhaps maddest)[6] logician in modern times, used *Principia Mathematica* as a basis to devise his incompleteness theorem (Gödel 1992). As described in Chapter 1, his theorem proves that there are arithmetic propositions which are neither or solvable or unsolvable (i.e. they are *undecidable*) within the arithmetic system. This indicates that no system can prove its consistency within its own framework. Metaphors for Gödel's Incompleteness theorem include trying to figure out for yourself whether or not you are insane, or trying to see your face with your eyes. This relates to the theory of logical types, which specifies that any statement about a collection involves all the collection, and must therefore not be a part of it. Because Gödel's theorem is concerned with provability within an arithmetical system, it is meta-arithmetical, and does not belong to mathematics,

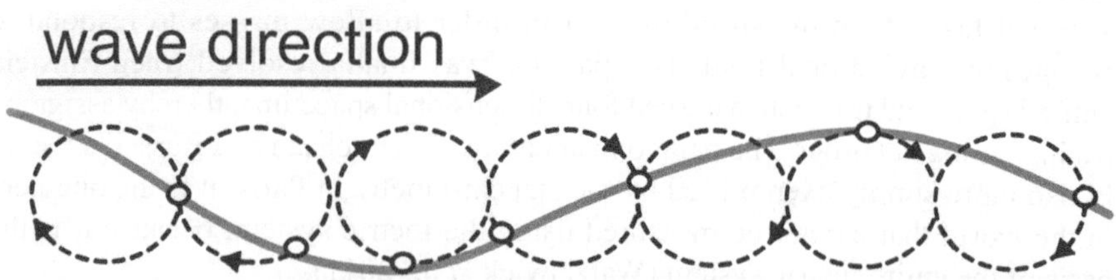

Figure 3.1 An illustration of how a surface wave (smooth line) is created by the circular clock-wise movements (hatched circles) of water particles (white dots)
Source: adapted from Capra (2010).

but rather to *meta-mathematics* (Braithwaite 1992). Thus, the theorem conjures an infinite series of Chinese boxes:

> You might be able to prove every conceivable statement about numbers within a system by going *outside* the system in order to come up with new rules and axioms, but by doing so you'll only create a larger system with its own unprovable statements. The implication is that *all* logical systems of any complexity are, by definition, incomplete; each of them contains, at any given time, more true statements than it can possibly prove according to its own defining set of rules.
>
> (Jones and Wilson 2009: 548)

Conclusion

Galois's group theory helps unravel Heraclitus's ancient paradox of the unity of opposites, demonstrating how it is that change within a system does not change the overall system itself, and rather reinforces the system whole instead. The theory of logical types demonstrates how system wholes are qualitatively (logically) different from their parts, and how paradox is generated when logical levels of phenomena are confused. While group theory is a way of thinking about change that occurs *within* invariant systems, the theory of logical types is a way of thinking about the "peculiar metamorphosis" that occurs when one *jumps* from one logical level to another, escaping one invariant system to enter another (Watzlawick *et al.* 2011: 11). This relates to Gödel's incompleteness theorem, which demonstrates that logical systems always contain a 'hole', or a piece of missing information. It is this 'hole' which creates the requirement for a qualitative 'jump' when moving from logical level or type to another. The critical importance of this notion will become apparent in Chapter 4, where I introduce the mysterious and paradoxical worlds of chaos and complexity.

Notes

1 Translated by D. W. Graham (2010: 159) from Heraclitus 69[F40].

2 In logic, vicious circles are circular errors in reasoning caused when the premise is used to prove the conclusion, and the conclusion used to prove the premise. In every day usage, vicious circles refer to situations where an apparent solution to a problem results in a chain of events that creates a new problem and increases the difficulty of solving the original problem (e.g. see *American Heritage Dictionary of the English Language*, www.ahdictionary.com/word/search.html?q=vicious+circle&submit.x=-857&submit.y=-210). From a systems perspective, vicious circles are systems with positive or escalating gain (Bateson 2002).

3 Average velocity can be calculated using algebra. Instantaneous velocity, however, requires calculus, specifically a derivative of position. Acceleration and jerk are calculated using second and third derivatives of position, respectively.

4 Bateson (2000) points out that Newtonian equations used to describe the motion of particles 'stop' at acceleration, as they are unable to account for the deformation of moving bodies required for change in acceleration to occur.

5 See Pennsylvania State University's wave animations at www.smartsciencepro.com/2-types-waves/.

6 Gödel died of malnutrition due to a phobia of being poisoned.

References

Bateson, G. (2000). *Steps to an Ecology of Mind: Collected Essays in Anthropology, Psychiatry, Evolution, and Epistemology*. Chicago, IL: University of Chicago Press.

Bateson, G. (2002). *Mind and Nature: A Necessary Unity*. Cresskill, NJ: Hampton Press.

Braithwaite, R. B. (1992). Introduction. In K. Gödel, *On Formally Undecidable Propositions of Principia Mathematica and Related Systems*. New York: Dover.

Cantini, A. (2012). Paradoxes and contemporary logic. *Stanford Encyclopedia of Philosophy* (Fall 2014 edition). Retrieved from http://plato.stanford.edu/entries/paradoxes-contemporary-logic.

Capra, F. (2010). *The Tao of Physics*. Boston, MA: Shambhala.

Cave, P. (2005). Humour and paradox laid bare. *The Monist*, 135–153.

Clement, K. C. (2005). Russell's paradox. *Internet Encyclopedia of Philosophy*. Retrieved from www.iep.utm.edu/par-russ.

Dell, P. F. (1981). Paradox redux. *Journal of Marital and Family Therapy 7*(2), 127–134.

Einstein, A. and Calaprice, A. (2005). *The New Quotable Einstein*. Princeton, NJ: Princeton University Press.

Fitzpatrick, S. (2005). Open-ended tangled hierarchies: Zen koans and paradox in public administration. *Intl Journal of Public Administration 28*(11–12), 957–971.

Gödel, K. (1992). *On Formally Undecidable Propositions of Principia Mathematica and Related Systems* (trans. B. Meltzer). New York: Dover.

Graham, D. W. (ed.). (2010). *The Texts of Early Greek Philosophy: The Complete Fragments and Selected Testimonies of the Major Presocratics*. Cambridge: Cambridge University Press.

Grof, S. (1981). Nature, mind, and consciousness: Gregory Bateson and the new paradigm. *Journal of Transpersonal Psychology 5/2/31*, 72.

Irvine, A. D. (2010). Principia Mathematica. *Stanford Encyclopedia of Philosophy* (Winter 2010 edition). Retrieved from http://plato.stanford.edu/cgi-bin/encyclopedia/archinfo.cgi?entry=principia-mathematica.

Jones, J. and Wilson, W. (2009). *An Incomplete Education: 3,684 Things You Should Have Learned but Probably Didn't*. New York: Random House.

Marx, G. (1959). *Groucho and Me*. New York: Bernard Geis Associates.

Northrop, E. P. (2014). *Riddles in Mathematics: A Book of Paradoxes*. New York: Dover.

Slater, B. H. (2005). Logical paradoxes. *Internet Encyclopedia of Philosophy*. Retrieved from http://www.iep.utm.edu/par-log.

Tarrant, J. (2005). Paradox, breakthrough, and the Zen koan. *Shift: At the Frontiers of Consciousness 6*, 24–27.

Watzlawick, P., Beavin, J. H. and Jackson, D. D. (1967). *Pragmatics of Human Communication: A Study of Interactional Patterns, Pathologies and Paradoxes*. New York: W. W. Norton.

Watzlawick, P., Weakland, J. H. and Fisch, R. (2011). *Change: Principles of Problem Formulation and Problem Resolution*. New York: W. W. Norton.

Whitehead, A. N. and Russell, B. (1913). *Principia Mathematica*. Cambridge: Cambridge University Press.

4 Filling the ‘hole’

Non-linearity, chaos and complex self-organising systems

> Unfortunately, non-chaotic systems are very nearly as scarce as hen’s teeth, despite the fact that our physical understanding of nature is largely based on their study … Algorithmic complexity theory and non-linear dynamics together establish the fact that determinism reigns only over a quite finite domain; outside this small haven of order lies a largely uncharted, vast wasteland of chaos.
>
> (Joseph Ford, ‘How random is a coin toss?’)

Linear to non-linear

In the 1920s, discoveries in the field of quantum mechanics demonstrated the profound interconnectivity of the universe. This revived interest in metaphysical holism among scientists and philosophers, including early ecologists and those in the organicism movement. However, by the time the dust of the Second World War had settled, these theories, with their panpsychic and idealist leanings, were lost to mainstream science, having succumbed to the purges of the logical positivists and their remnants swallowed by war. Newtonian reductionism made its forceful revival in the scientific realism that came to dominate scientific thinking of the post-war era. Yet despite its spectacular successes, reductionist science, which breaks everything down into its simpler parts, leaves a vacuum in its lop-sidedness:

> How do we use the information gleaned from the parts to build up a theory of the whole? The deep difficulty here lies in the fact that the complex whole may exhibit properties that are not readily explained by understanding the parts.
>
> (Kauffman 1995: vii–viii)

The origin of these inexplicable properties is perhaps the ‘hole’, or missing information, in the reductionist system of thought. If so, it cannot be understood within the framework of Newtonian reductionism, as per Gödel’s incompleteness theorem and the theory of logical types. Rather, filling the ‘hole’ involves radically reconceptualising the ‘whole’.

One of the first clues as to what might fill this 'hole' in our understanding of reality had appeared decades earlier, in 1887, when a major advance was made in the field of non-linear mechanics by the French polymath Henri Poincaré. Briggs and Peat (1989) note that Newton's linear equations had enabled scientists and engineers to relate rates of change to various forces and create the technological marvels of the fledging industrial age. They were, however, unable to explain non-linear phenomena, such as explosions, high winds, and sudden breaks in materials. Non-linear behaviour was poorly understood, and even the simplest non-linear equations of the time were too difficult to solve in most cases. Capra (1997) describes how natural non-linear phenomena, such as air or water turbulence, were viewed as too chaotic to contend with, and were generally avoided by scientists. When confronted with non-linear problems, engineers would 'linearise' them, replacing them with linear approximations, a type of differential equation that served to hide system complexities. Application of linear approximations became so habitual that many engineers and scientists came to believe that most natural phenomena behaved in linear ways (i.e. simply and predictably) despite most natural systems and all living systems being "relentlessly non-linear" and complex in reality (Stewart 1989, quoted in Capra 1997: 122). Capra argues that this view, which reinforced reductionism and denied complexity, took hold in the in 1900s and persisted for most of the twentieth century.

Poincaré and topology

Which leads to the relevance of Poincaré's discovery and the science of *topology*, which are described by Capra (1997) and Briggs and Peat (1989) as follows. In the 1600s, the focus of mathematics had swung from the more qualitative image-based geometry, to quantitative analysis via formula-based algebra. Whereas solutions from linear equations can be generalised to other problems, this is not the case with non-linear equations, which tend to be peculiar. As such, non-linear systems are better suited to qualitative (descriptive), rather than quantitative (numerical) analysis. In the late 1800s, Poincaré devised the science of topology as a way of qualitatively analysing the incredibly complicated behaviours of non-linear problems. Topology (otherwise known as *rubbersheet geometry*) is a geometry whereby lengths, angles, areas, etc. can be continuously distorted such that they transform into different shapes; as such, it is the study of continuity and connectivity. Topologically equivalent figures can be distorted into each other. For example, a triangle can be distorted into a rectangle, the rectangle into an octagon, then octagon into a circle. These figures are topologically equivalent. Likewise, a pyramid can be distorted into a cylinder; therefore the two are topologically equivalent. However, a square cannot be distorted into a cube; as such they are *not* topological equivalents. Topology discerns those properties that are not changed when a figure is distorted. For example, when a square is distorted into a circle, it remains two-dimensional. When a cube is distorted into a cylinder it remains three-dimensional. The holes in a shape are also invariant—a ball can never be transformed into a doughnut and vice versa; however, a doughnut is topologically equivalent to coffee mug. It is these invariant properties (or patterns) that prevent topological non-equivalents from transforming into each other; thus topology is a mathematics of "relationships, of unchangeable, or 'invariant' patterns" (Capra 1997: 126). Topology is linked to logic theories, such as those described in Chapter 3, in that both explore the relationship between the continuous and the discrete (Stillwell 1993). Poincaré used topology as a way of graphically 'mapping' the behaviours of complex systems of in an abstract mathematical space called *phase space*. This enabled him to understand the global geometry of systems of differential equations (Florin 2000).

Poincaré was interested in the workings of Newtonian mechanics in closed systems. At the time, closed systems in physics (e.g. a pendulum swinging in a vacuum) were regarded as perfectly orderly and predictable, as per the Newtonian model of the universe. Any unpredictable, random and chaotic behaviour in observed systems was attributed to disturbance or 'contamination' from unforeseen outside forces, indicating the system was not perfectly closed. Using Newton's laws, physicists were able to precisely predict the effects of two celestial bodies interacting in a hypothetical system. However, once a third body was added to the system, Newton's equations became unsolvable, and an exact solution could no longer be worked out. Using a method called perturbation theory, physicists would instead apply a series of successive approximations to try and converge on an answer they hoped would be correct. Poincaré had observed that the approximation method worked well for the first few terms of the three-body system, but was curious about what would happen to the planet's orbits over the course of millions of years. To work this out, Poincaré added a term to the two-body equation reflecting the feedback effect caused by adding the third body, thereby increasing the non-linearity of the equation. When he applied his topological analysis it revealed a profound complexity, one he could not attempt to draw. Poincaré had 'seen' the hallmark of chaos: a *strange attractor*. Poincaré found that even if the third body caused only a very minute perturbation, with time some orbits would become erratic and cause planets to wobble—sometimes so much so that planets would be flung out of the solar system. He had inadvertently stumbled across the effects of *chaos*: an intrinsic feature of non-linear systems whereby tiny effects can be radically amplified into destabilising and transformative forces. His findings implied that if left sealed in a 'box' indefinitely (to guard against external disturbance), a completely determinate non-linear system could, at any time, spontaneously develop its own instabilities and become chaotic. This discovery so disturbed Poincaré that he wrote: "these ideas are so bizarre that I cannot bear to contemplate them" (quoted in Briggs and Peat 1989: 29). At the close of the nineteenth century Poincaré abandoned his "bizarre ideas", and it was not until almost a century had passed that they were revisited.

Non-linear mathematics

Despite making strides in management, engineering and cognitive sciences, by the 1970s systems thinking was either ignored or viewed as a failure by mainstream sciences. This was evinced in Robert Lilienfeld's critique of systems theory, in which he states: "No evidence that systems theory has been used to achieve the solution of any substantive problem in any field whatsoever has appeared" (Lilienfeld 1978: 192). Capra attributes this view to the fact that general system theory had not been developed into a mathematical discipline. Mathematics of the time was still largely limited to linear equations that were unable to describe complex, non-linear phenomena (i.e. most of nature). Indeed, engineers and mathematicians had been masking complex phenomena for decades via the use of linear approximations (Capra 1997). Briggs and Peat (1989: 23) describe non-linear equations as a mathematical twilight zone where "Solvers making their way through an apparently normal mathematical landscape can suddenly find themselves in an alternate reality". They explain the nature of such equations as follows. In linear equations, small changes lead to small effects and big changes lead to big effects. In simple non-linear equations, however, one small change has the potential to generate huge, sometimes catastrophic changes, thanks to the power of feedback. A key feature of non-linear equations is that they incorporate feedback, meaning A causes B, then B has an effect on A. This creates a feedback loop, which is a form of iteration, whereby the system repeatedly works back on itself. Iteration and negative feedback (whereby B's effect is to limit A) can stabilise a system, as in the case of homeostasis. Iteration and positive feedback (whereby B's effect is to amplify A), however, can cause a system to 'run away', as in the case of explosions. Iteration is typical of non-linear systems, and is responsible for the rich complexity often seen within them. It also makes a system inherently unpredictable, even in the case of strictly deterministic equations (where all the rules and variables remain constant). A non-linear system can exhibit a constant pattern of behaviour for long periods, then suddenly, at a critical point, change its behaviour altogether or devolve into seemingly random, chaotic behaviour. Non-linear equations are too complex to be solved using analytic methods (manipulating an equation until you get a final solution). Instead numerical methods are required, involving lengthy trial-and-error search for a combination of variable numbers that 'fits' the equation. Resolving such problems by hand or with early computers was impossible.

In the 1960s, Poincaré's topological system was revived as a way of visualising non-linear systems that were too difficult to describe using mathematical computations and complex non-linear equations. Mathematicians learned they could bend and twist topological shapes to represent how a system moves or behaves. Early chaos theorists began using topology to map the behaviour of non-linear phenomena and in doing so caught some of science's first glimpses of strange attractors, patterns of implicit order that emerge when non-linear systems become chaotic. By the late 1960s, investigations into non-linear systems had intensified with the formulation of non-linear mathematics and the advent of high-speed computers. According to Capra, the discovery of non-linear mathematics is possibly one of the most important events of the twentieth century. It propelled the construction of new mathematical frameworks describing the behaviour of dynamic non-linear systems, such as *dynamical systems theory* and one of its main branches, *chaos theory*. Because these frameworks describe the global nature of systems, they were soon being used to explain an array of phenomena across scientific disciplines, including turbulence, weather, lightning strikes, blood vessel networks, traffic patterns, stock market fluctuations, plant growth, and many more (Gleick 1987). Non-linear mathematics were also used to forge research into self-organising and *complex systems*--systems having global behaviour that cannot be explained solely by looking at their constituent parts.[1] Models of complex self-organising systems requiring computation of thousands of inter-dependent feedback interactions were finally made possible by the use of increasingly powerful computers in the 1970s and 1980s (Capra 1997).

This gave rise to a variety of complexity sciences and theories (for an overview see Wells 2013), aiming to describe the "dynamic interactions of multiple elements engaged in self-organising processes"[2] in an expanding range of physical, chemical, biological, social and computerised systems (Wells 2013: 50). As touched on in Chapter 2, synergetics, hypercycles and Gaia theory were among them, as were non-linear thermodynamics, autopoiesis, chaos theory and complexity theory. In this chapter, I provide a simple sketch of selected concepts from these latter theories, drawing on the works of Prigogine and Stengers (1984), Briggs and Peat (1989), Capra (1997), Maturana and Varela (1980), Holland (1995) and Kauffman (1993). From this I hope to illustrate how chaos, complexity and self-organisation relate to phenomena of a higher logical order than those explained by the reductionist Newtonian paradigm, thereby proffering a different set of premises for understanding and thinking about reality. As Gleick writes:

> Where chaos begins, classical science stops. For as long as the world has had physicists inquiring into the laws of nature, it has suffered a special ignorance about disorder in the atmosphere, in the turbulent sea, in the fluctuations of wildlife populations, in the oscillations of the heart and the brain. The irregular side of nature, the discontinuous and erratic side—these have been puzzles to science, or worse, monstrosities.
>
> (Gleick 1987: 3)

Dissipative structures and non-linear thermodynamics

The most important early description of self-organising systems was that of Belgian chemist, physicist and Nobel Laureate Ilya Prigogine, who was driven by an early intuition that life's order might somehow spring from friction and energy. Prigogine began studying non-equilibrium thermodynamics, what was then a small field concerned with the thermodynamics of systems in continuous flux and therefore not in thermodynamic equilibrium (i.e. most natural systems; Briggs and Peat 1989). Like the much more prominent discipline of molecular biology, this field had sprung from Erwin Schrödinger's 1944 work *What is Life*. Specifically, it built on Schrödinger's proposal that living things are open systems that 'drink orderliness' from their environments, concentrating it on themselves as a means of replicating themselves and escaping entropic decay into disorder. According to

Briggs and Peat, it took a long time for Prigogine's understanding to break through, as he was, in his words, "a prisoner of the *linear* non-equilibrium theory" (Briggs and Peat 1989: 139), and erroneously trying to find answers by looking at systems modelled using linear approximations. Then, in the 1960s, Prigogine made the critical discovery that systems far from equilibrium—which include all living systems—must be described with non-linear equations (Capra 1997).

To help build this understanding of living systems, he turned to simpler physical phenomena that also self-organise in conditions far from equilibrium: the Benard instabilities[3] of heat convection (see Nicolis and Prigogine 1977; Prigogine 1967; Prigogine and Lefever 1968). Benard instabilities are strangely ordered patterns that emerge when a thin layer of liquid is uniformly heated from below. Capra (1997) describes their formation as follows. As the liquid is heated, heat travels up via conduction, or flux (the heat moves up but the liquid molecules stay mostly still). This is a near equilibrium situation. However, as the liquid is further heated, and temperature difference between the upper and lower layers of liquid grows, the system moves far from equilibrium. Gravity pulls harder on the upper layer, which is cooler and therefore denser. The fluid begins to move in increasingly turbulent whorls and vortices, as cooler parts of the fluid sink and warmer parts rise. At this point, the system is chaotically dissipating heat through both conduction and through the movement of fluid molecules. Once the temperature difference between the top and the bottom of the liquid reaches a threshold value, the system arrives at a critical point of instability. At this time, a stable pattern of hexagonal cells, known as Benard cells, spontaneously forms in the liquid. This signals the point where the system abandons heat conduction in favour of orderly *convection*, whereby heat is dissipated via coherent motion of the liquid's molecules. The hexagonal pattern is created as hot liquid moves up the centre of the Benard cells, and cooler liquid descends down the cell walls to the bottom. With continued heating, however, the hexagonal cells dissolve back into turbulent chaos. Prigogine and Stengers (1984: 142) describe Benard cells as a "spectacular phenomenon" whereby millions of molecules suddenly begin moving together in a coherent pattern.

Thus, pushing a system far away from equilibrium (e.g. by heating it up in the case of Benard cells) to the point that it becomes turbulent (or chaotic) can result in the spontaneous formation of order via self-organisation. Prigogine described these formations as *dissipative structures* because they paradoxically create structure and order while dissipating energy. As Capra (1997) describes it, the increasing flow of energy and matter through a dissipative structure creates amplifying positive feedback loops, which generate the critical instability (turbulence) that causes the system to *jump* to a new form or level of organisation. Prigogine's work indicates that dissipative structures not only maintain themselves in stable states far from equilibrium, they also *evolve*. With increasing through-flow of energy and matter, dissipative structures may pass through additional points of instability and stability, transforming into new, evermore complex structures. As such, dissipative structures give rise to irreversible processes. Other examples of non-living dissipative structures include lasers, hurricanes, whirlpools and autocatalytic chemical cycles (hypercycles).

Through his work Prigogine developed the important field *non-linear thermodynamics* to describe phenomena that lead to self-organisation (i.e. increased order) in *open systems far from equilibrium*. This is in contrast to classical thermodynamics, which describes the inevitable loss of order/organisation in *closed systems close to equilibrium*. The key distinction here is that in classical thermodynamics energy dissipation creates 'waste' in the form of heat and increases entropy, whereas in non-linear thermodynamics energy dissipation creates order and reduces entropy. Prigogine also developed a new understanding of the term 'chaos'. In science the term has traditionally been used to describe what Prigogine calls *equilibrium thermal chaos*—the "passive chaos of equilibrium and maximum entropy, where elements are so intimately mixed that no organization exists" (Briggs and Peat 1989: 136). Prigogine brought into parlance *far-from-equilibrium turbulent chaos*, which Briggs and Peat (*ibid.*) describe as hot, active and energetic—an active rather than passive dissipation of organisation.

Autopoiesis

Another important description of self-organisation in living systems originated from the work of Chilean biologist Humberto Maturana. Maturana had a keen interest in system organisation spurred by his collaborations with cyberneticists in the 1950s, including Warren McCulloch (his PhD supervisor) and Walter Pitts. He did not believe that the organisation of living systems could be explained by enumerating their properties (Maturana and Varela 1980). Indeed, he believed that over-emphasis on isolated components of living systems by molecular, genetic and evolutionary biology was the greatest hindrance to understanding life (Varela *et al.* 1974). While agreeing with the organicist notion that organisms needed to be understood as unities, or wholes, Maturana felt the organicist approach failed to adequately explain the relationship between the whole and its parts, and how the whole arises (Maturana and Varela 1980). In the 1960s, while working as a neurologist, he combined his interest in the 'organisation of the living' with his questions around what happens during the phenomenon of perception. He formalised his ideas with his student Francisco Varela, and in 1972 they introduced the concept of *autopoiesis* to describe the self-organising nature of living systems (*ibid.*), what Cohen and Wartofsky (1980) call a topological theory of biology. Rather than concerning themselves with the question "What are the necessary properties of the components that make a living system possible?" they asked "What is the necessary and sufficient organization for a given system to be a living unity?" (Varela *et al.* 1974: 197).

Autopoiesis is premised on the notion that living entities (i.e. organisms) are actually *units of interaction*, as opposed to 'things', calling to mind the *actual occasions* of Alfred Whitehead's process philosophy. These units of interaction are generated by *circular organisation*, the source of order in all living systems. Circular organisation constitutes a system's *identity,* and emerges from the interactions of the system's component entities[4] (which themselves are units of interaction). These components, whose continuous interactions together 'create' the emergent system whole, must be constantly produced and maintained by the system whole (via downward causation), such that their interactions continue to produce the system whole (via upward causation). This combination of upward and downward causation maintains the circular organisation of the whole. Thus, there is no division between *producer* and *product* (Maturana and Varela 1992): they are mutually interdependent. As an example, a cell (the whole) is generated by interactions between various biochemical components, such as nucleic acids and proteins; in turn, the cell works to continuously produce and maintain these biochemical components such that they continue to produce the cell. Hence, the function of an organism's circular organisation is simply to maintain its circular organisation, and thus its *identity,* in a continuously changing environment. Circular organisation continuously brings the organism back to the same internal state—it is the source of homeostasis (and hence the organism). It creates a *self-referring*, or self-sustaining, system—an *autopoietic unit*, "a factory that makes itself from within" (Luisi 2003: 52). This is in contrast to an *allopoietic* system, such as a machine, which through its functioning produces something other than itself, such as a car or a car factory. Thus, an organism is a complex system, an emergent autopoietic unit which maintains its identity "as long as the basic circularity that defines it as a unit of interactions remains unbroken" (Maturana and Varela 1980: 9—11). In autopoiesis, boundaries[5]—such as cell walls, membranes, skin, etc.—specify the domain of the network's operations and define the system as a unit. This differs from hypercycles (Chapter 2), which are self-organising but have no boundaries, and are regarded as precursory rather than actual living systems (Capra 1997).

Autopoiesis explains how living systems maintain their essential identity, or overall pattern, while their parts continuously interact and change, reflecting the paradoxical interdependence between persistence and change demonstrated by Galois's group theory (Chapter 3). As Maturana and Varela state, living systems are "organised in a closed causal circular process that allows for evolutionary change in the way the circularity is maintained, but not for the loss of the circularity itself" (Maturana and Varela 1980: 9). Although organisationally closed by virtue of their circular processes, living systems are physically open—dissipative structures, maintained by a continuous through-flow of energy and matter, as per Prigogine's non-linear thermodynamics. It is this inflow and dissipation of energy and nutrients that drives autopoiesis. As an example, Milo and Phillips (2014) note that our stomach cells are replaced every two to nine days, and those of our lungs every eight days. Yet despite this constant flux, these organs maintain their pattern of circularity and their resultant identity.

While Maturana and Varela describe the autopoietic unit as biologically autonomous, they also describe how it is paradoxically wedded to its environment. From the autopoietic perspective, an organism cannot be understood independently of its niche, defined as the entire set (or domain) of interactions into which the organism may enter, a notion that will be explored in greater detail in Chapter 6. Nor can the niche be understood separately from the organism—they co-emerge together. As long as they do not disrupt one another's circular organisation (which depends on their ranges of tolerance), autopoietic units can interact with each other without losing their identity. If their interactions become strongly coupled, autopoietic units will harmonise into higher and increasingly complex orders of unity, which are also self-referring. The component autopoietic units then become subservient to the maintenance of these higher order entities (Maturana and Varela 1980). In a human being, for example, cellular interactions couple to form tissue, tissues couple to form organs, and organs couple to form the whole human body. All these levels of autopoietic organisation are mutually interdependent: human cells can't survive outside the whole body, and the body can't survive without its component cells. Each level of autopoietic organisation behaves in a qualitatively different way to its component parts (e.g. you can't infer a human just by looking at its cells), reflecting the hierarchies of logical levels denoted by the theory of logical types.

The theory of autopoiesis has evolved significantly over time. It has been extended to a wide range of biological, chemical, and social systems, and has had particular influence in the fields of cognitive science and family therapy (see Luisi 2003). According to Luisi (*ibid.*), autopoiesis is the only theory that provides a unified view of the living, from the molecular level to human perception.[6] Nonetheless, it has had relatively little impact on mainstream biological science, a situation Luisi ascribes to the theory's 1970s origination, a time when biology was "completely dominated by a vision of DNA and RNA as the holy grail of life," (*ibid.*: 58).

Chaos theory

As Prigogine, Haken and others were empirically discovering that self-organising systems become chaotic and turbulent when pushed beyond certain thresholds, a similar breakthrough was afoot in the world of mathematics. In the 1950s, American meteorologist Edward Lorenz began to question the effectiveness of linear statistical models being used to forecast weather, a non-linear phenomenon. In 1963, he pioneered the beginnings of chaos theory by using non-linear differential equations to describe turbulent flows. In doing so he founded one of the central precepts of chaos theory: that an imperceptibly small change in the state of a non-linear system could cause a large difference in its future state. Consequently, if "there is any error whatever in observing the present state—and in any real system such errors seem inevitable—an acceptable prediction of an instantaneous state in the distant future may well be impossible" (Lorenz 1963: 133).

From his solutions, Lorenz concluded that precise long-range weather forecasting is impossible. This conclusion was also reflected in his 1972 paper 'Predictability: Does the flap of a butterfly's wings in Brazil set off a Tornado in Texas?' (Lorenz 1972), from which the term 'butterfly effect' was coined. In the 1970s and 1980s, Lorenz's work and the arrival of high-speed computers led to a growing number of mathematicians studying chaos in dynamical[7] non-linear systems. They found that deterministic[8] systems subjected to iterative feedback become chaotic and unpredictable (indeterminate) when pushed beyond critical boundaries, just as Poincaré did while studying the three-body problem in 1887.

Chaos theorists took interest in how *control parameters* from outside a system drive the system behaviours to one state or another. In addition to finding that small changes can lead to unpredictably large effects in these systems, they also found that simple non-linear systems can behave in complex ways, and complex non-linear systems can behave in simple ways. This was in direct contrast to linear systems, where small changes lead to small predictable effects, simple systems behave in simple ways, and complicated systems behave in complicated ways.

Phase space

Gaining a basic understanding of chaos theory requires an introduction to the concept of *phase space*. As discussed earlier, the behaviours of non-linear phenomena are not easily described using quantitative methods involving equations. Rather, it is easier to use qualitative methods to visualise their behaviours. Phase space is a qualitative technique for graphically viewing system behaviour in abstract mathematical space. Using this technique, every dimension, or variable, of a system's behaviour is represented by a different coordinate on an axis in abstract space. Briggs and Peat (1989) use the simple example of a three-dimensional object moving through space, such as an airplane flying over the Atlantic. The airplane has a phase space consisting of the three spatial dimensions, plus its speed in the direction of any one of these dimensions—hence it has a six dimensional phase space. The system's phase space represents the total range of behaviours the system (i.e. the airplane) can potentially 'explore'. It also delineates the boundaries of the system's possible behaviours. Thus phase space is a representation of both the system's total possibilities and its outer limits. When the actual trajectory of the airplane is mapped onto the system's phase space, it occupies only a tiny fraction of the phase space[9] available to the plane (reflecting that the plane has traversed just a small portion of the air space available to it).

Attractors

Stable orderly systems, such as the airplane described above, inhabit a small subset of their phase space. In chaos theory, the subset of phase space to which a system is repeatedly attracted is referred to as the system's *attractor*.[10] Attractors are like valleys or basins on a landscape—everything tends to be drawn into them. There are different types of attractors, which are described by Briggs and Peat (1989) as follows. When a system tends to return to the same fixed state or position after being disturbed, this state is called a *point* or *fixed attractor*. A simple example of a system with a point attractor is a pendulum. However you 'perturb' the pendulum (by pushing it one way or another), and however furiously it swings, friction causes the pendulum to come to rest at the exact point where it started: the bottom of its trajectory. When mapped in phase space, the behaviour of a system governed by a point attractor looks like a one-dimensional point (Figure 4.1a).[11]

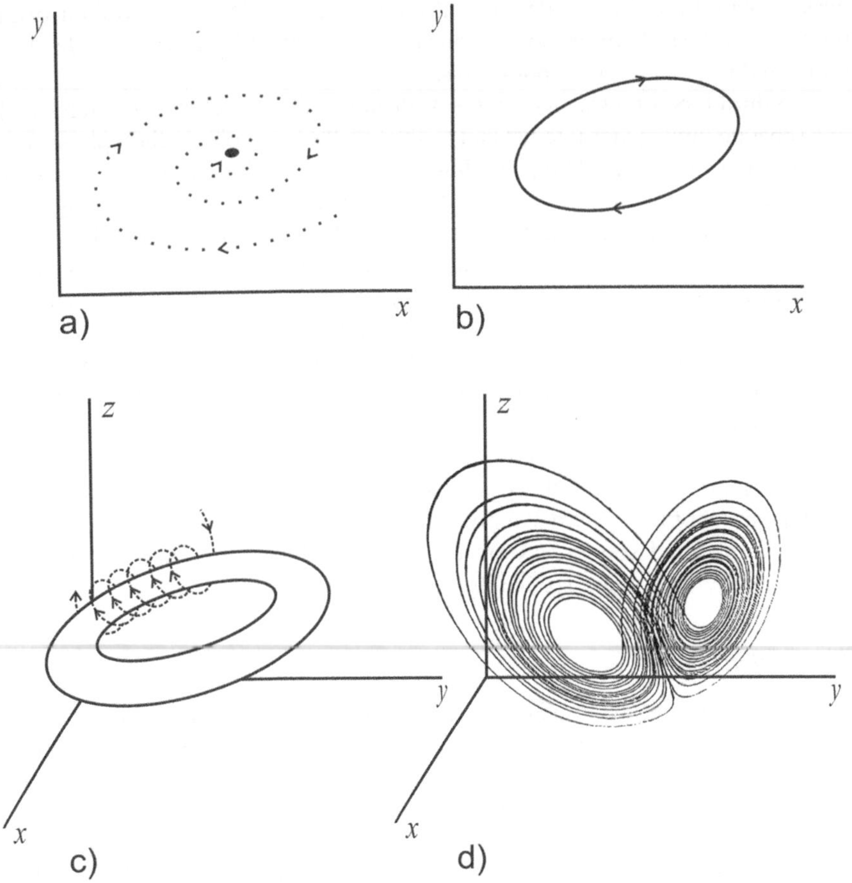

Figure 4.1 (a) Point, (b) limit cycle, (c) torus and (d) strange attractors as they appear when mapped in phase space

The second type of attractor is a *limit cycle*. Systems under the influence of a limit cycle oscillate between two different point attractors, like two valleys separated by a saddle. A pendulum given a periodic kick to overcome friction so it has a regular swing is an example of a system with a limit cycle attractor. The pendulum oscillates between two positions, or attractors, representing the height of its trajectory on either side of its arc. When mapped in phase space, the trajectory of a system governed by a limit cycle traces a two-dimensional circle (Figure 4.1b). Systems with more than two variables form more complex shapes, as their phase space will have more dimensions. The system is attracted to this cyclic path, rather than to just a fixed point. The regular cycling of predator–prey populations (e.g. trout and pike) through peaks and troughs is another classic example of a system governed by a limit cycle. Systems governed by limit cycles are remarkably stable. Even when seriously perturbed, say the predator or prey population is killed off by disease, with time the population will recover and settle back into their original limit cycle. The state of the population can be conceptualised as a point that moves round and round a circle in phase space.

The third type of attractor is a *torus attractor*. A torus attractor emerges when two separate limit cycles interact and fasten together, conjoining their phase spaces. An example would be if two pendulums were conjoined such that the swinging of one affected the swinging of the other. Another would be the joining of two predator–prey cycles, such as trout and pike with insects and frogs. When mapped in phase space, the trajectory of a system governed by a torus attractor assumes the shape of a three-dimensional doughnut. Again, this is assuming that each of the conjoined limit cycles only consist of two variables. If there are additional variables, or if more systems are conjoined, the torus will assume a much more complicated shape. The behaviour of the conjoined system can be visualised as a point looping around and around on the surface of the torus (Figure 4.1c). The system is attracted to this looping path, rather than just a cyclic path. Systems with torus attractors are quite orderly and have *asymptotic* predictability: although the exact position of the system at a given time is unknown, it can still be certain that it will be situated somewhere on the surface of the torus, and not randomly wandering through phase space (Briggs and Peat 1989).

Bifurcations

Non-linear systems are attracted to stable attractors, such as point, limit cycle, and torus attractors, when their control parameters are set at low levels (i.e. 'flowing' slowly through the system). Control parameters are external inputs to a system that, as they are changed, move the system through different attractor states. A control parameter might be the rate of energy or information flow, or population growth. As Stacey (1996) explains, when these parameters are set at low levels, the instabilities they generate are generally small enough to be damped down by the systems' negative feedback loops. However, as these parameters are 'dialled up' (so they flow faster through the system), the instabilities they generate begin to exceed the system's damping capacity, as small disturbances to the system are amplified exponentially via positive feedback. At a critical threshold, the behaviour of the system begins to *bifurcate* and its stable pattern becomes more complicated.

Briggs and Peat (1989) describe a bifurcation as a point of instability in a system where its behaviour branches off in a qualitatively different direction, forming a new attractor. To illustrate this effect, they use the example of a non-linearly growing population as shown in Figure 4.2, with growth rate charted against the population as a percent of its original size. The behaviour of the population, or system, starts out being governed by a point attractor. As its growth rate increases, the system reaches a critical point[12] ($\sim r = 3.0$) where it bifurcates and suddenly starts oscillating between two stable values, or states. The system's behaviour is now governed by a limit cycle attractor. If the growth rate continues to increase, the system reaches yet another bifurcation point ($\sim r = 3.5$), and now the population starts oscillating between four stable values. The system is now governed by a

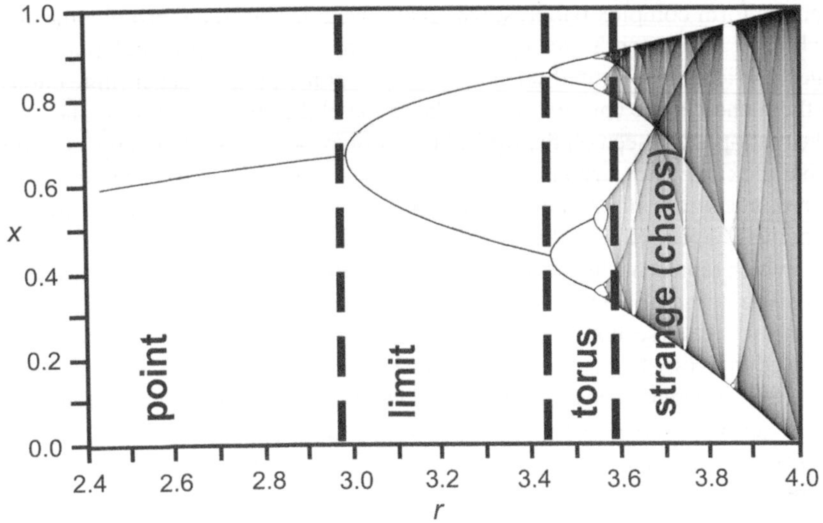

Figure 4.2 Graph of a non-linearly growing population, with growth rate (r) charted against the population (x) as a per cent of its original size. As the growth rate increases, the system trajectory repeatedly bifurcates, causing it to transition from a point attractor through to a strange attractor

torus attractor. With more rate increases, the system quickly bifurcates again and again, oscillating between eight stable values, then sixteen. Although still predictable, the system is now very complicated; in phase space the surface of the system's torus attractor is beginning to break apart. From this point, the system's behaviour rapidly descends into chaos: jumping wildly and unpredictably around in phase space.[13] The surface of the system's torus attractor has now broken up and entered a *fractional dimension*, whereupon it may then form a *strange attractor* (e.g. Figure 4.1d).

Strange attractors

In the left and centre areas of Figure 4.2 (where $r = 2.4$ to 3.6), the trajectory of the system occupies a small subset of its phase space. In contrast, the darkened areas of the graph (where $r = 3.6$ to 4.0) show the system chaotically jumping all over, filling up its phase space. However, one can also see an underlying *meta-pattern* in the chaos: the arching black lines where there is the highest probability of finding the system, and the intermittent white bands—interludes of stability where the system oscillates between fixed points again. As the system iterates, it bends and folds on itself, creating bands of order and chaos (Briggs and Peat 1989). When its chaotic behaviour is mapped, it is shown to occupy a completely new region of phase space. Although the system's behaviours never quite repeat themselves while in the chaotic phase, they follow *similar* trajectories. In abstract phase space, their iterations form complex patterns, which are irregular but recognisable, the shape of which is the system's *strange attractor* (*ibid.*). The system's behaviour is no longer restricted to a limited selection of stable points: it now has infinite choices. But these choices are confined within the shape of the strange attractor, the source of the meta-pattern seen on the right side of Figure 4.2. As such, strange attractors are paradoxical: they are simultaneously stable and unstable, ordered and disordered, at equilibrium and disequilibrium (Stacey 1996). Strange attractors take distinctive shapes, such as the well-known Lorenz (or butterfly) attractor (Figure 4.3a), which represents the behaviour of the Lorenz weather system in phase space; the Rossler attractor (Figure 4.3b), representing a simplified model of the Lorenz system; or the horseshoe attractor, discovered in the 1960s by a mathematician who used topology to examine noise in an electrical feedback system (Briggs and Peat 1989). Whereas chaos was once considered equivalent to randomness or entropy, it is now understood that chaotic systems actually have deep, underlying patterns of order that assume qualitatively distinct shapes—strange attractors (Capra 1997).

Dimensions of order

To help the layperson understand chaos, Briggs and Peat (1989) use an interesting metaphor. Think of a stable system as a piece of paper, a two-dimensional plane. Then start crumpling it—the folds and crumples represent turbulence, or chaos.

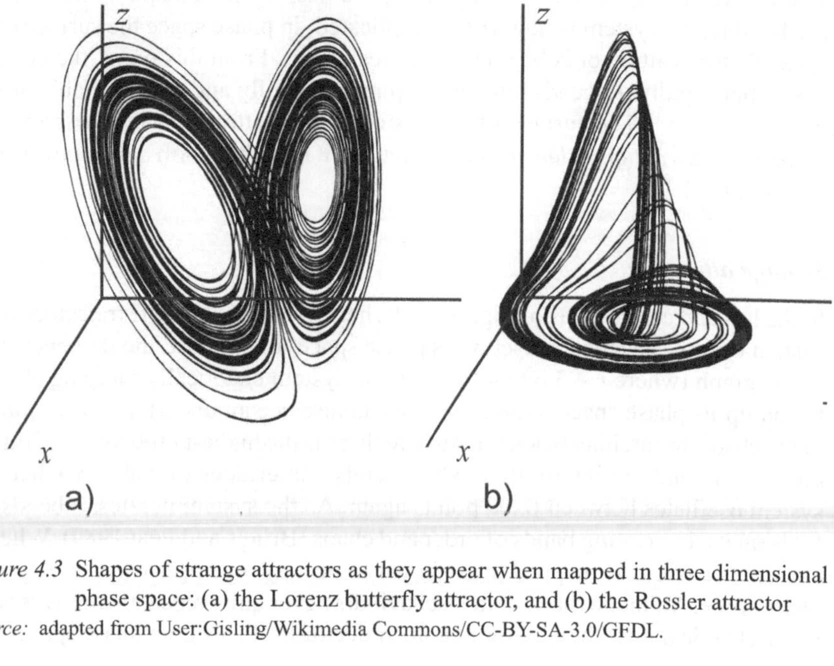

Figure 4.3 Shapes of strange attractors as they appear when mapped in three dimensional phase space: (a) the Lorenz butterfly attractor, and (b) the Rossler attractor

Source: adapted from User:Gisling/Wikimedia Commons/CC-BY-SA-3.0/GFDL.

The more you crumple the paper together the more it begins to resemble a three-dimensional ball. The paper is now *caught* between being a two-dimensional plane and a three-dimensional solid. Chaos indicates a *transition* where the system is caught between different dimensions, i.e. *qualitatively* different system states or levels of order. Once it transforms from two dimensions to three, the new system state is no longer *topologically equivalent*[14] to the previous. A system approaching this transition point is often referred to as being *on the edge of chaos*. As a system approaches the edge of chaos, it becomes highly sensitive: even the slightest perturbation can amplify into radical changes in its behaviour.

The transitioning of a system through increasing levels of order toward chaos has an odd parallel with the relationship between position, velocity and acceleration discussed in connection to the theory of logical types in Chapter 3. Using calculus, velocity can be calculated as a first order derivative of position. Acceleration is calculated as a second-order derivative of position, and jerk (a change in acceleration) as a third-order derivative of position. Interestingly, non-linear systems described by jerk equations may begin to show chaotic behaviour, and increasingly so in those described by *hyperjerk* (involving fourth-order plus derivatives) equations (Chlouverakis and Sprott 2006). Prigogine's experiments with Benard cells also illustrate this process. As the fluid is heated, it begins to move convectively. When modelled simply in phase space the system's trajectory changes from a point attractor to a limit cycle. With continued heating the fluid reaches a critical point where turbulence, or chaos, sets in. This represents a point of indecision for the system, where it fluctuates wildly between a higher and lower dimension of order (a conductive vs. a convective system), until it finally jumps to the higher dimension: a stable hexagonal pattern of Benard cells. When mapped in phase space, the system trajectory now forms a strange attractor.

Like Gödel's theorem, strange attractors exhibit "Chinese boxes of subtle order", inhabiting the fractional realm that "exists in the cracks of things … between the first, second and third dimensions of the familiar world with its point attractors, limit cycles, and well-run tori" (Briggs and Peat 1989, citing David Ruelle: 89). Briggs and Peat (1989) illustrate this using the fascinating example of Californian researchers who recorded the time intervals between 4000 drips of a faucet, looking for order in what was long assumed to be a random system. When graphed as points in phase space the drip intervals made a decidedly non-random pattern. In fact, the pattern was a cross section of a strange attractor previously generated by an iterating equation developed by a French scientist named Michel Henon. When the researchers zoomed in on these patterns, they found they were self-similar, smaller scale versions of the same pattern. When they increased the faucet's water pressure and repeated the experiment, the graphed patterns formed cross-sections of new, previously unknown strange attractors.

Fractals

Fractal geometry is another means of describing chaotic patterns. Capra (1997) provides a description as follows. Developed in the 1960s by the brilliant mathematician Benoit Mandelbrot, fractal geometry explains the irregular but self-similar patterns that are seen everywhere in nature, patterns which cannot be described with conventional geometry (Figure 4.4). Fractal geometry was soon taken up by mathematicians to describe the fine-scale structure of chaotic patterns. Although fractal patterns are often incredibly complex, they are described using very simple iterative mathematical formulas (i.e. formulas that repeat back on themselves, creating feedback loops). For example, the Mandelbrot set of mathematical points is generated by iterating the complex quadratic polynomial $z_{n+1} = z_n^2 + c$. Surprisingly, mapping the points generated by this simple formula creates the most complex mathematical object ever invented (Figure 4.5), one of bewildering, bewitching and almost indescribable complexity. The Mandelbrot set is best experienced in animated format, which allows the viewer to 'zoom through' the object's complex boundary.[15]

The irregularity, or jaggedness, of fractal patterns is an expression of system's chaos. Fractal patterns repeat themselves at descending scales, such that the pattern of the whole is reflected in the pattern of the parts. Schneider (2009, citing others) notes that Mandelbrot's work disrupts the view of organisms as Euclidean objects with two-dimensional surfaces living in three-dimensional environments. Rather, organisms have convoluted surfaces that are fractal, and live in environments that are likewise fractal. Fractal patterns are found everywhere in nature: plants, forests, coastlines, snowflakes, mountains, clouds, blood vessels, river systems, and seashells.

Fractal patterns also illustrate the limitations of quantitative analyses. For example, Mandelbrot (1967) related fractal geometry to the question "How long is the coast of Britain?" Because the coast is jagged, and this jaggedness is self-similar at all scales, it depends on the scale at which you measure it (Figure 4.6).

Figure 4.4 Trees are examples of natural fractal objects. Notice the branching repeats itself at different scales, but is never exactly the same

Figure 4.5 A partial view of the fractal pattern generated by the Mandelbrot set
Source: Wolfgang Beyer/Wikimedia Commons/CC-BY-SA-3.0/GFDL.

Figure 4.6 How long is the coast of Britain? Because the coast is jagged, and this jaggedness is self-similar at all scales, it depends on the scale at which you measure it (Mandelbrot 1967). The coast is indefinitely long if measured with smaller and smaller units

Source: Alexandre Van de Sande Wikimedia Commons/CC-BY-SA-3.0/GFDL.

The coast is indefinitely long if measured at smaller and smaller scales. In other words, quantities of fractal (chaotic) systems can never be definitively defined because they are dependent on the scale of observation. However, the coast can be described *qualitatively,* by generating a ratio of its relative jaggedness. Jaggedness is a measure of the system's chaos (the more jagged, the higher the fractal dimension and therefore more chaotic) and is relative to scale.

Sensitivity to initial conditions and the edge of chaos

There is another key feature of dynamic non-linear systems that highlights the limits of quantitative analysis: their extreme sensitivity to initial conditions. Because these systems are iterative, very tiny perturbations can be amplified into very large effects through positive feedback.[16] Briggs and Peat (1989) illustrate using measurement as an example. Measurement has finite precision and computers can only do finite calculations. Therefore infinite strings of decimal places must be rounded off at some point. Yet no matter how many decimal places are conserved in the numbers fed into non-linear equations, after a number of iterations the round-off error is amplified to such a degree that the computations become chaotic and prediction is no longer possible. Gregory Chaitin (1982) used a new information theory proof to show Gödel's incompleteness theorem is not just a 'mathematical curiosity'. Rather, it is a reflection of this iterative paradox, with Gödel's missing information corresponding to the uncertainties and rounding-off errors in non-linear equations that are amplified through iteration. This missing information sows the seeds of chaos and unpredictability found in most natural systems.

As such, even when modelling deterministic non-linear systems, which in theory should be predictable, chaos inevitably emerges and their behaviour becomes paradoxically *indeterminate*. Prigogine and Stengers (1984) note that non-linear systems are not only sensitive to fluctuations produced by their own internal activity, but to fluctuations from their external environment as well. Crutchfield *et al.* (1986) provide a vivid illustration of this extreme sensitivity. They state that if one was able to measure and control every single variable involved in striking a billiard ball, the trajectory of the ball could no longer be predicted after one minute if one's calculation neglected the gravitational effect of a single electron on the edge of the galaxy:

> The large growth in uncertainty comes about because the balls are curved, and small differences at the point of impact are amplified with each collision. The amplification is exponential: it is compounded at every collision, like the successive reproduction of bacteria with unlimited space and food. Any effect, no matter how small, quickly reaches macroscopic proportions. That is one of the basic properties of chaos.
>
> (Crutchfield *et al.* 1986: 50)

This vast sensitivity is indicative of a system's interconnectedness, its wholeness. For this reason, complex systems are resistant to reductionist analysis requiring some part of the system to be cut off from 'external' feedback loops (i.e. holding other parts of the system constant) so it can be examined in isolation. This is compounded when scientists go the reductionist route of attempting to measure and model parts of dynamic non-linear systems: the round-off error or the 'missing information' quickly renders prediction impossible. Uncertainty and errors with regard to the system's initial state seed the system's future with potential turbulence and chaos (Briggs and Peat 1989). Hence, order and chaos co-exist within non-linear systems.

Causality

The Newtonian conception of a determinate universe is one where all events are the result of cause and effect (upward causation). Therefore, given complete information about all the forces at work in nature, one can predict the future, as famously conjectured by Laplace in what Polyani (1962: 141) called the "Laplacian Delusion". This notion of the universe assumes that all processes are reversible, and therefore the universe is timeless in that there is no *arrow of time.* Thus the irreversibility we perceive in our day-to-day lives (e.g. heat conduction, diffusion, chemical reactions, thought, etc.) is actually an illusion. Newtonian physicists noticed the irreversible effects of friction but ignored them. Nineteenth century thermodynamics put a dent in the concept of reversible time with the introduction of entropy, but this was clawed back when Boltzmann devised probability theory to reconcile entropy with Newtonian laws. Much of quantum theory landed on the side of classical mechanics by likewise assuming the reversibility of time.[17] It was finally Prigogine, through his work on open systems far from equilibrium, who presented strong (albeit controversial) arguments in favour of an arrow of time (Prigogine *et al.* 1969; Prigogine and Stengers 1984; Prigogine and Antoniou 1999). Prigogine argues that the dynamic bifurcations of complex systems reveal that time is irreversible, pointing to the irreducible (and therefore irreversible) complexity emerging out of chaos in dissipative systems. He states that while causality is a continuous force, each bifurcation of the system is a decision-point where many futures are possible. At such points, the system seems to hesitate among its various possibilities. The path the system then chooses is unpredictable, a matter of chance rather than causality. And once it chooses the other potential futures vanish irreversibly.

Creativity

Prigogine also notes that bifurcation points involve amplification of small effects. He regards the amplification capacity of self-organising systems to be their "creative lever" (Briggs and Peat 1989: 145), as it allows for the spontaneous emergence of novel patterns of order. As discussed earlier, when a system is approaching chaotic, or turbulent, conditions (i.e. the edge of chaos), it becomes highly sensitive to small disturbances, which can potentially blow up into large, unexpected effects. As such, Briggs and Peat note that biological systems employ negative feedback to dampen and prevent amplification of small effects and maintain homeostasis (i.e. their ordered pattern). However, the reverse is true in parts of biological systems requiring flexible and creative behaviour; they maintain edge-of-chaos conditions, and are therefore very responsive to small effects and external stimuli (*ibid.*). This effect has special relevance to the complexity sciences and the development of complexity theory.

Complexity theory

Despite apparent commonalities with cybernetics and systems theory, the complexity sciences emerged separately in the 1990s as a result of computer-based explorations into non-linear dynamics as well as interest in genetics and biological evolution (Heylighen *et al.* 2007). In the 1990s, John Holland, a computer scientist and engineer, elucidated his ideas on complexity in a series of lectures for the Santa Fe institute, a non-profit centre dedicated to the interdisciplinary study of complex systems. The institute was set up in 1984 by scientists previously working with the Los Alamos National Laboratory who were disenchanted with highly reductionist and specialised approaches to science, and who were intrigued by recent advances in non-linear dynamics and computer modelling (Dillon 2001). Holland later compiled these lectures into his book *Hidden Order How Adaptation Builds Complexity* (Holland 1995), which he followed with *Emergence* in 1998 (Holland 1998). Expanding from his work on genetic algorithms, Holland developed computer simulations to investigate *complex adaptive systems* (also referred to as *multi-agent systems*), a term coined by the Santa Fe Institute to describe complex systems that are capable of learning and adapting to their surroundings (see also Gell-Mann 1995; Waldrop 1994). Complexity theorists are mainly concerned with patterns of spontaneous order that emerge from the self-organising interactions of networked entities, and the behaviour of ordered systems nearing the edge of chaos (as per Langton 1986). Chaos theorists narrowly study the dynamics of non-linear systems using deterministic mathematical models. In contrast, complexity theorists are more interested in understanding the dynamics of non-linearity in the 'wild', in a range of natural and artificial systems, from galaxies and ant colonies to political economies and artificial intelligence (see Brownlee 2007 for an overview of contributions). For a good summary of the attributes of self-organising systems and principles of self-organisation see Cilliers (1998).

Adaptation and emergence

Complex adaptive systems are dynamic networks comprised of large numbers of interacting entities referred to as *agents* by complexity scientists. These agents are constantly acting and reacting to what the other agents in the system are doing (i.e. they are adapting). Holland (1995) indicates that this notion of adaptation expands the biological usage of the term (a process whereby an organism fits itself to its environment to make better use of it over time) to include learning-based processes, which he equates to information processing. Living systems are "the paramount examples" of complexity (Kauffman 1993: 173): cells, organisms, ecosystems, bird flocks and cities are all examples of complex adaptive systems. The overall behaviour of the system is the result of a huge number of decisions made every moment by many individual agents, without any external or centralised control or design. Computer modelling shows that as multiple independent agents continuously interact, they spontaneously organise and reorganise themselves into more elaborate structures over time via feedback. Once formed, the self-organised emergent order typically resists change.[18] Complexity scientists refer to the phenomena of large-scale order, or patterns, emerging from small-scale interactions between individual parts, or agents, as *emergence*. Because the whole of the complex system is greater than the sum of the parts, the system will exhibit emergent patterns of behaviour not seen in its constituent agents on their own.[19]

Holland (1987) believes that complex adaptive systems interact with their environments in 'game-like' ways, meaning they are motivated by pay-offs that afford them with continued existence and adaptation. The environment supplies an enormous range of niches, a subset of which is exploited by the complex adaptive system. According to Holland, there is a trade-off between exploration and exploitation of these niches. A complex adaptive system can exploit an old niche where there are well-established payoffs, or it can explore new possibilities in a new niche, but at the cost of leaving the old niche and its predictable pay-off for unknown gains. He describes emergence largely in relation to game-theory, as patterns that emerge from game playing agents (Holland 1998), which is reflective of models applied by classical economists (e.g. wage-contract games).

Boolean or binary networks

Another early complexity scientist with the Santa Fe Institute was evolutionary biologist Stuart Kauffman. In his work, Kauffman (1993) used binary, or Boolean, networks developed by cyberneticists in the 1940s to model enormously complex biological and chemical systems that could not be described using differential equations. Capra (1997) provides a basic description of how these binary networks operate using the simple binary networks illustrated in Figure 4.7. These networks each consist of six nodes, with each node connected to three neighbouring nodes. Each node is capable of having a value of either ON (coloured black) or OFF (coloured white). Simple switching rules can be established for the network, whereby the value of a node is determined by the prior values of its neighbouring nodes. Figure 4.7 illustrates sequences generated by the following rule: if at least two of its neighbours are ON at the start of the sequence, a node will be ON in the next step; otherwise it will be OFF. This creates sequences of on-off states within the nodes, and different patterns of sequences can be seen to emerge as the network cycles. These sequences can then be mapped in six dimensional phase space, and classified according to their attractors. For example, the sequence may settle into a stable state where the nodes are all off, and hence the network no-longer changes (i.e. a point attractor; sequence A in Figure 4.7). Or the sequence might run a few steps then start oscillating between two different states (i.e. a limit cycle attractor; sequence B in Figure 4.7). Because binary networks are deterministic[20] (the network has a finite number of states, regardless of how large it is), given enough time they settle into one of their attractors and remain in that state.

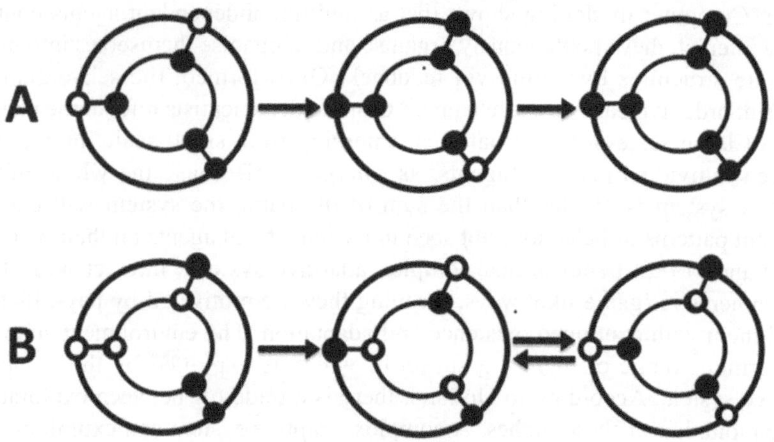

Figure 4.7 Two sequences of binary network states generated by a simple ON/OFF rule. By the third step, sequence A settles into a stable state, or point attractor, with all nodes switched off. Sequence B settles into a limit cycle, where it oscillates between two states

Source: adapted from Capra (1997).

NK model and the edge of chaos

Kaufmann (1993) examined the behaviours of many large binary networks, comprised of millions of interacting variables, to try and discern the relationship between chaos and order in complex systems. He discovered that these "massively disorganised" (*ibid.*: 173) networks spontaneously self-organise to coordinate the activities of thousands and even billions of elements, generating what he calls *order for free*. From his observations Kauffman noted that the behaviours of these networks fall into three broad regimes: chaotic, ordered and complex. The behaviour of the binary networks is driven by two key parameters: the number of nodes (N) in the network, and the average number of connections (K) between the nodes. If the number of connections diminishes, the system eventually freezes into a fixed state of activity such that its behaviours have become highly ordered and unchanging—it enters an *ordered regime*. These systems *crystallise* order, such that changes cannot propagate through the network. Such ordered states arise when most of the system elements fall into either active or inactive states. They have small, ordered attractors, boxed into localised regions of phase space

In cases where the number of connections between nodes increases, the system eventually enters a *chaotic regime*. Here the system fluctuates erratically between multiple states. The attractor[21] is very large, often infinitely long, and an exponential function of the binary variables in the system consisting of billions of seemingly random states. (Note that this understanding of chaos as a state of disorganisation differs from that of chaos theorists, who use the term to describe deep underlying patterns of order generated by determinate systems driven past certain thresholds, and the order generating *far-from-equilibrium turbulent chaos* described by Prigogine. Kauffman's use of the term is more reflective of the turbulence that occurs before a dissipative system jumps to a more highly ordered state of turbulent chaos.) In the chaotic regime, the system is very sensitive to initial conditions and small changes, which can cause large *avalanches* of change to propagate through it. As the system moves from an ordered to a chaotic state, it passes through a transition phase where the system is spanned by a "connected, percolating, *frozen*" core of nodes, which remain unchanged as the rest of the system fluctuates in complex ways (Kauffman 1993: 234). In this phase small changes are likely to create *small* avalanches of change, and occasionally large avalanches. Kauffman called this transition phase[22] between ordered and complex the *edge of chaos*, the *complex regime*.[23] From his NK models, Kauffman hypothesised that living systems characteristically occupy this half-way state on the edge of chaos, where they have sufficient flexibility to adapt and evolve while still maintaining stability in the essential processes that give them their identity. Deeply ordered or chaotic systems, in contrast, are probably not highly evolvable or capable of complex behaviours.

Fitness landscapes and evolution

With respect to evolution, Kauffman sees an inseparable relationship between selection and self-organisation. An organism alters its biophysical environment while being simultaneously altered by the same environment. The dynamic of a system is co-determined by its own internal dynamic, and by the size and connectivity of the other systems that are part of the bigger system to which it also belongs. Thus all organisms are coupled with their environments, cooperating and competing among themselves, and co-evolving in dynamically shifting *fitness landscapes*.[24] He goes on to suggest that linked co-evolving natural systems are mutually moved "as though by an invisible hand" (Kauffman 1993: 29) to the edge of chaos. Here they form ecosystems through which co-evolutionary changes propagate, with areas of instability propagating both speciation and extinction. When Kauffman ran NK fitness landscape models, he found that fitness landscapes with higher K values (i.e. connections between nodes) sustained higher overall fitness, with species co-evolving more quickly into sustained fitness and Nash equilibrium.[25]

Kauffman (1995) also used his evolution simulation models to show that randomly interacting entities in a system very quickly connect via feedback loops into autocatalytic strings. These strings then form autocatalytic networks (hypercycles), all in absence of any selective pressure. From this he was able to demonstrate that individual protein molecules could organise themselves into self-reproducing assemblages of enzymes. Kauffman's findings relate back to biochemical research conducted by Manfred Eigen, who discovered self-organising processes in biochemical systems far from equilibrium. Eigen and his colleagues found catalytic enzymatic reactions exposed to energy flows would self-organise into complex networks of closed loops, or *catalytic cycles* (Eigen and Schuster 1977), in a process Capra (1997) summarises as follows. With continued energy flow, over time catalytic cycles will successively self-organise into more and more complex structures called *hypercycles*. Each new level of organisation occurs after the system passes through a successive period of instability. These cycles play an essential role in the metabolic function of living things, and are so stable that they can persist under a wide range of conditions. They are also capable of self-replication and correction of replication errors. As such, hypercycles are considered to be precursors to living systems. Eigen and Schuster suggest that this process whereby a new, complex structure emerges from a period of instability can also be applied to evolution and mutation.

In the Darwinian understanding of evolution, new life forms gradually evolve as useful variations accumulate under selective pressures. While he does not completely discount Darwinian evolution, Kauffman (1993) proposes that the 'order for free' exhibited by self-organising systems helps explain the emergence of life and its full array of complexity. Although complex systems are under continuous selective pressure, Kauffman's work shows that the more complex that systems are, the more resistant they are to alteration as a result of selective pressures. His work also shows that complex systems spontaneously self-organise into "profound order" (*ibid.*: 29). This new order is maintained *if* selective pressures *fail* to squash it. Thus, Kauffman conjectures that spontaneous self-organisation is evolution's driving force,[26] with new life forms emerging *in spite* of selection, not because of it. He states, "Evolution is not just 'chance caught on a wing.' It is not just a tinkering of the ad hoc, of bricolage, of contraption. It is emergent order honoured and honed by selection" (*ibid.*: 644). This relates to recent work by the physicist Jeremy England (2013) who has devised a new mathematical formula to explain how energy applied to groups of atoms causes them to arrange into increasingly complex structures and/or become self-replicating as they work to dissipate energy. England proposes that dissipative processes may inexorably cause atoms to take on the attributes of life, and may indeed be the underlying drivers of evolution.

Kauffman's models have proven to have predictive value in a number of areas in molecular biology. Capra points out that the various attractors of binary systems represent the behaviour of a large number of natural systems, including ecosystems, immune systems, and genetic systems. For example, it has been shown that an organism's genome operates as a vast network of interconnected feedback loops, whereby genes regulate each other's activities, a notion that differs substantially from earlier views of a genetic code acting like a "biochemical computer executing a 'genetic program'" (Capra 1997: 199).

Creative wholeness versus reductionism

Kauffman (1993: 235) emphatically states that "*our intuitions about the requirements for order in very complex systems have been wrong*". He argues that the modern reductionist paradigm begins to fail once it attempts to describe the self-organising natural (and human) world (Kauffman 1995). This is because the complex wholes of these systems, governed by their own emergent laws or properties, cannot be explained by examining their parts. Indeed, he sees theories of emergence as pointing to our wholeness with the universe, as we participate in its collective unfolding. He also sees the endlessly creative phenomena of emergence as a vehicle for 'reinventing the sacred' in the modern world (*ibid.*).

Although Holland (1998) argues that emergence of rule-governed systems is the opposite of reduction, his approach to understanding complex adaptive systems, unlike Kauffman's, nonetheless emanates strongly from the paradigm of mechanism and reductionism. Holland's work focuses on decomposing complex phenomena into their component parts and determining their causal mechanisms in aid of building models that can improve prediction and control of such systems. In his descriptions of how agents adapt (e.g. Holland 1995), he uses terms such as: performance, inputs, outputs, messages, signals, processing, performance, and syntax. In doing so, Holland reveals a conception of organisms as information processing machines, or computers. Indeed, he defines agents as "a collection of message-processing rules" (*ibid.*: 46). As such, his theory is *cognitive*, meaning it relies on system agents having internal models of external reality which in turn govern the agents' behaviour (the implications of which are detailed in Chapter 6). Holland argues that understanding how complex adaptive systems work will help us locate lever points in the system, and therefore resolve problems in such systems (e.g. immune disease, inner city decay, business innovation, etc.). In his view, an understanding of emergence will help us control previously uncontrollable aspects of the world, by allowing us to find ways of fostering the emergent phenomena we want, and discouraging the ones we do not (Holland 1998).

Conclusion

By the end of the Second World War, holism had been stripped from mainstream science and replaced with a renewed emphasis on reductionism. Heylighen *et al.* (2007) suggest that reductionism eclipsed holism because Newtonian approaches were so successful that it was thought they could explain everything. By comparison, holism had a mystical sensibility that seemed to lack scientific foundation (*ibid.*). Holism pointed to the tendency of wholes to be greater than the sum of their parts, but prompted the question, what exactly is it that is 'more' in the whole (*ibid.*)? From the above discussion, we see this question has now been answered by the chaos and complexity sciences, fields of study concerned with the emergence of structure, order and patterns from complex and apparently disordered/chaotic systems. Kauffman (1995: 69) describes his complexity theories as "rooted in unrepentant holism, born not of mysticism, but of mathematical necessity". And so it is that mathematics have filled the vacuum left by reductionism's domination of mainstream science. Smuts's holism was referring to the phenomenon complexity scientists call *emergence* (Heylighen *et al.* 2007), *emergent properties* being the 'more' in the whole—the qualitatively novel patterns of behaviour that spontaneously self-organise in non-linear systems operating far from equilibrium. Emergence is the critical piece of information missing from reductionist thinking, as per Gödel's Incompleteness theorem, and the source of the qualitative 'jump' that takes a system from one logical level to another, as the greater whole emerges from the interactions of its parts. In furnishing this missing information, chaos and complexity theory have revealed holism and reductionism to be paradoxically wedded in a unity of opposites, as per the ancient teachings of Heraclitus.

Franzen (2005) argues that invoking Gödel's theorem outside the realms of math and logic is inappropriate. Gödel's theorem has, however, been legitimately shown as:

- a means of mathematically explaining emergence in complex systems (e.g. Baas and Emmeche 1997); and
- a source of (logical) complexity in unpredictable human systems.

(Biggiero 2001)

Chaitin's information theory "suggests that the incompleteness phenomenon discovered by Gödel is natural and widespread rather than pathological and unusual" (Chaitin 1982: 941), indicating the *premise* behind Gödel's theorems is more than just an isolated paradox. Gödel's icompleteness theorem suggests that intractable problems cannot be solved using the rules and premises governing the system within which the problems are situated.[27] They can only be solved by adding new premises or 'rules', by 'jumping' to a higher logical level and into a new frame of reference, as per the theory of logical types. The central question of this book relates to why knowledge transfer and adaptive management so often fail to improve the generation, sharing and application of knowledge for environmental management. In the chapters that follow, I will show how this problem is deeply rooted in the metaphysical assumptions of the Newtonian paradigm, and how the complexity sciences can help furnish new premises and a new paradigm for reframing it.

Notes

1 Although this is a frequently used definition, what defines a complex system is contentious and some complexity scientists refute anti-reductionist interpretations of complexity; see Horgan (1995).

2 Wells (2013: 2) notes that while complexity is often seen as a feature exclusive to dynamic systems, she believes complexity is a "profound characteristic of reality" and characteristic of *all* systems, a view I agree with.

3 Also referred to as Raleigh–Benard convection.

4 As Maturana and Varela (1980: 49) state "the notion of *component* is necessary only for epistemological reasons in order to refer to the genesis of organization through our *description*, but this use does not reflect the nature of its composition".

5 Maturana and Varela were able to produce autopoietic patterns and boundaries using cellular automata (Varela *et al.* 1974). In doing so, they were among the first scientists to simulate self-organising systems. Cellular automata are mathematical network models developed subsequent to McCulloch and Pitt's cybernetic binary models, and used as alternatives to differential equations for modelling complex systems (see Von Neumann and Burks 1966).

6 Autopoiesis is not without its detractors. Birch (1991), for example, claims Maturana's work to be unoriginal, flawed, and even absurd.

7 A dynamical system is a mathematical model that has a rule specifying how the system changes over time.

8 A deterministic system has fixed rules and no randomness, and therefore should theoretically always produce the same output (i.e. it should be predictable). A dynamical system is deterministic.

9 Note this conception of phase space is a gross simplification of the actual phase space available to the airplane. If one considers that the plane consists of particles, the actual phase space would be 6*n*, where *n* is the number of particles that comprise the plane (Briggs and Peat 1989). Examination of other variables, such as acceleration or deceleration, requires adding even more dimensions to the plane's phase space.

10 This relates to Ashby's (1962) early cybernetic principle that dynamical systems evolve through self-organisation towards a state of *equilibrium,* which is now understood as an *attractor.*

11 See Briggs and Peat (1989: 35–41), for more detailed graphical descriptions of attractors mapped in phase space.

12 Scientists have found that that these critical bifurcation points occur at the point where the system's period (i.e. the time it takes for the system to oscillate back to its starting point) doubles (Briggs and Peat 1989).

13 Another easy to visualise example is the effect of increasing stream flow around a rock. Under low flow, water parts smoothly around the rock. As the flow increases, a stable vortex forms behind the rock, which then bifurcates into more and more vortices. Eventually, the vortices erupt into complete turbulence, and there is no longer any apparent order: the flow has become chaotic (Briggs and Peat 1989).

14 Note that none of the attractors in Figure 4.1 are topologically equivalent—they each successively represent the system having jumped to a higher, qualitatively different level of order.

15 E.g. see www.youtube.com/watch?v=ivRQDbAduoM.

16 Briggs and Peat (1989) describe how a system's susceptibility to chaotic behaviour depends on whether it is periodic or quasi-periodic. A system is periodic when the start and end points of its behavioural cycle are the same (this happens when periods/frequencies of the two coupled systems form a simple ratio, a rational number). A system is quasi-periodic when it looks periodic but never exactly repeats itself, so its end point is always different from its starting point (this happens when the periods/frequencies of the two coupled systems do not form a ratio; rather they are irrational, having decimal expression with infinite, non-repeating terms). Quasi-periodic systems are more stable because they never return to the same point between oscillations, therefore there is little opportunity for resonance (which creates amplification and positive feedback, and can potentially knock the system out of its pattern/cycle of behaviour). Periodic systems, on the other hand, are unstable because they always return to the same point between oscillations, making them prone to resonance with repeated iterations, i.e. amplification/positive feedback. Because of this, periodic systems are susceptible to descending into chaos and "gothic complexity" (*ibid.*: 43). This observation has been backed by observations of Jupiter's asteroid belt. Gaps have been found in the places where an asteroid's orbit would be in a periodic cycle with Jupiter's orbit, indicating that any asteroids attempting to occupy the gaps are likely to be shot off into space (*ibid.*).

17 Although nineteenth-century understandings of entropy and Darwinian evolution led to scientific acceptance of irreversibility at the macroscopic scale, the microscopic laws of both classical and quantum mechanics have long been considered reversible because they are believed to be *time symmetric*, meaning they remain unchanged whether running forward or backward in time (i.e. there is no distinction between past and future). Prigogine controversially proposed that the laws of physics should be reformulated to introduce irreversibility as their basis (see Prigogine and Antoniou 1999). Prigogine's time symmetry breaking ideas were criticised for lacking empirical and mathematical support. Recent research, however, has begun to suggest microscopic foundations for irreversibility and the arrow of time in quantum settings (e.g. Batalhao *et al.* 2015; Silva *et al.* 2014).

18 This relates to Lovelock's (1991) mathematical computer models of the earth's self-regulating Gaian system. These models reveal that when the feedback cycles linking living and non-living aspects of the earth's systems are broken, populations of organisms begin to fluctuate wildly and the system becomes chaotic. As the complexity of the feedback networks in the model are increased, the planet's self-regulation becomes more stable, with an improved ability to recover from severe disturbance.

19 This is equivalent to the *step function* in cybernetics.

20 As formal deterministic models, binary networks have well-defined variables to ensure predictability and computability.

21 Kauffman (1995) describes strange attractors as having different *dimensionalities* of order, depending on how much of their phase space they use. Whereas high dimensional strange attractors wander through vast tracts of phase space, low dimensional strange attractors box their behaviour into relatively small portions of their available phase space (though not as small as point, limit and torus attractors), and are much more ordered as a result. As such, the chaotic regime as described here would be equivalent to a *high-dimensional* strange attractor.

22 This relates to Bak and Chen's (1991) theory of self-organised criticality, whereby a self-organising system is drawn to a critical point between order and chaos, optimising as many attractor states as it can without becoming unstable.
23 The complex regime being equivalent to a *low-dimensional* strange attractor.
24 Fitness landscapes are models used for visualising problems of optimisation or adaptation; high fitness forms peaks on the 'landscape' and low fitness forms valleys.
25 A level of stability reached once no player, or agent, in the system will gain further benefit by altering its strategies.
26 While Kauffman refutes Darwinian selection as a source of complexity and instead sees complexity arising spontaneously as a result of systems moving towards chaos, Holland's thinking is more Darwinian—he views adaptation as the source of complexity.
27 See also Cilliers's (1998) notion of incompressibility in complex systems.

References

Ashby, W. R. (1962). Principles of the self-organizing system. In H. v. Foerster and G. W. Zopf (eds), *Principles of Self-Organization: Transactions of the University of Illinois Symposium* (pp. 255–278). London: Pergamon Press.

Baas, N. A. and Emmeche, C. (1997). On emergence and explanation. *Intellectica 25*(2), 67–83.

Bak, P. and Chen, K. (1991). Self-organized criticality. *Scientific American 264*(1).

Batalhao, T., Souza, A., Sarthour, R., Oliveira, I., Paternostro, M., Lutz, E. *et al.* (2015). Irreversibility and the arrow of time in a quenched quantum system. *arXiv preprint* arXiv:1502.06704.

Biggiero, L. (2001). Sources of complexity in human systems. *Nonlinear Dynamics, Psychology, and Life Sciences 5*(1), 3–19.

Birch, J. (1991). Re-inventing the already punctured wheel: reflections on a seminar with Humberto Maturana. *Journal of Family Therapy 13*, 349–373.

Briggs, J. and Peat, F. D. (1989). *Turbulent Mirror: An Illustrated Guide to Chaos Theory and the Science of Wholeness*. New York: Harper & Row.

Brownlee, J. (2007). *Complex Adaptive Systems*. CIS Technical Report 070302A. Melbourne, Australia: Complex Intelligent Systems Laboratory, Swinburne University of Technology.

Capra, F. (1997). *The Web of Life*. London: Flamingo.

Chaitin, G. J. (1982). Goedel's theorem and information. *International Journal of Theoretical Physics 22*(12), 941–954.

Chlouverakis, K. E. and Sprott, J. (2006). Chaotic hyperjerk systems. *Chaos, Solitons and Fractals 28*(3), 739–746.

Cilliers, P. (1998). *Complexity and Postmodernism: Understanding Complex Systems*. London: Routledge.

Cohen, R. S. and Wartofsky, M. W. (1980). Editorial preface. In *Autopoiesis and Cognition: The Realization of the Living by Maturana, H. and Varela, F.* (pp. v–vi). Dordrecht: Reidel.

Crutchfield, J. P., Farmer, D., Packard, N. and Shaw, R. (1986). Chaos. *Scientific American 254*, 46–57.

Dillon, D. (2001). *A Review of the Santa Fe Institute: Institutional and Individual Qualities of Expert Interdisciplinary Work*. Cambridge, MA: Project Zero, Harvard Graduate School of Education.

Eigen, M. and Schuster, P. (1977). The hypercycle: A principle of natural self□organization. *Naturwissenschaften 64*, 541–565.

England, J. L. (2013). Statistical physics of self-replication. *Journal of Chemical Physics 139*(12), 121923.

Florin, N. (2000). The integral manifolds of the three body problem (English summary). In D. G. Babbitt and J. E. Kister (eds), *Featured Reviews in Mathematical Reviews 1997–1999: With Selected Reviews of Classic Books and Papers from 1940–1969*. Providence, RI: American Mathematical Society.

Ford, J. (1983). How random is a coin toss? *Physics Today 36*(4), 40–47.

Franzen, T. (2005). *Gödel's Theorem: An Incomplete Guide to Its Use and Abuse*. Wellesley, MA: Peters.

Gell-Mann, M. (1995). *The Quark and the Jaguar: Adventures in the Simple and the Complex*. New York: Henry Holt.

Gleick, J. (1987). *Chaos*. New York: Penguin Books.

Heylighen, F., Cilliers, P., and Gershenson, C. (2007). Complexity and philosophy. In J. Bogg and R. Geyer (eds), *Complexity, Science and Society* (pp. 117–134). New York: Radcliffe Publishing.

Holland, J. H. (1987). Complex adaptive systems: A primer. *Santa Fe Institute Bulletin 2*(1), 8–10.

Holland, J. H. (1995). *Hidden Order*. Reading, MA: Helix.

Holland, J. H. (1998). *Emergence*. Oxford: Oxford University Press.

Horgan, J. (1995). From complexity to perplexity. *Scientfic American 272*(6), 104–109.

Kauffman, S. (1993). *The Origins of Order: Self-Organization and Selection in Evolution*. New York: Oxford University Press.

Kauffman, S. (1995). *At Home in the Universe: The Search for Laws of Self Organization and Complexity*. New York: Oxford University Press.

Langton, C. G. (1986). Studying artificial life with cellular automata. *Physica D: Nonlinear Phenomena 22*(1–3): 120–149.

Lilienfeld, R. (1978). *The Rise of Systems Theory: An Ideological Analysis*. New York: John Wiley & Sons.

Lorenz, E. (1963). Deterministic nonperiodic flow. *Journal of the Atmospheric Sciences 20*(2), 130–141.

Lorenz, E. (1972). Does the flap of a butterfly's wings in Brazil set off a tornado in Texas? Paper presented before the American Academy for the Advancement of Science, 29 December.

Lovelock, J. E. (1991). *Healing Gaia: Practical Medicine for the Planet*. New York: Harmony.

Luisi, P. L. (2003). Autopoiesis: a review and a reappraisal. *Naturwissenschaften 90*(2), 49–59.

Mandelbrot, B. B. (1967). How long is the coast of Britain? *Science 156*, 636–638.

Maturana, H. and Varela, F. J. (1980). *Autopoiesis and Cognition: The Realization of the Living*. Dordrecht: Reidel.

Maturana, H. and Varela, F. J. (1992). *The Tree of Knowledge: the Biological Roots of Human Understanding*. Boston, MA: Shambhala.

Milo, R. and Phillips, R. (2014). *Cell Biology by the Numbers*. Retrieved from www.rpgroup.caltech.edu/courses/aph161/2012/files_2012/homework/CaltechJan2012.pdf.

Nicolis, G. and Prigogine, I. (1977). *Self Organization in Nonequilibrium Systems*. New York: Wiley.

Pagels, H. R. (1985). Is the irreversibility we see a fundamental property of nature? *Physics Today 38*(1), 97–99.

Polanyi, M. (1962). *Personal Knowledge: Towards a Post-Critical Philosophy*. London: Routledge & Kegan Paul.

Prigogine, I. (1967). Dissipative structures in chemical systems. In S. Claesson (ed.), *Fast Reactions and Primary Processes in Chemical Kinetics*. New York: Interscience.

Prigogine, I. and Antoniou, I. (1999). Laws of nature and time symmetry breaking. *Annals of the New York Academy of Sciences 879*(1), 8–28.

Prigogine, I. and Lefever, R. (1968). Symmetry breaking instabilities in dissipative systems. II. *The Journal of Chemical Physics 48*(4), 1695–1700.

Prigogine, I. and Stengers, I. (1984). *Order Out of Chaos: Man's New Dialogue with Nature*. New York: Bantam Books.

Schneider, D. C. (2009). *Quantitative Ecology: Measurement, Models and Scaling*. London: Academic Press.

Schrödinger, E. (1944). *What is Life? The Physical Aspect of the Living Cell*. Cambridge: Cambridge University Press.

Silva, J. P., Sarthour, R. S., Souza, A. M., Oliveira, I. S., Goold, J., Modi, K. *et al.* (2014). Experimental demonstration of information to energy conversion in a quantum system at the Landauer Limit. *arXiv preprint* arXiv:1412.6490.

Stacey, R. (1996). *Complexity and Creativity in Organizations*. San Francisco, CA: Berrett-Koehler.

Stewart, I. (1989). *Does God Play Dice? The New Mathematics of Chaos*. Oxford: Blackwell.

Stillwell, J. (1993). *Classical Topology and Combinatorial Group Theory* (Graduate Texts in Mathematics, vol. 72). Berlin: Springer.

Varela, F. G., Maturana, H. R. and Uribe, R. (1974). Autopoiesis: the organization of living systems, its characterization and a model. *Biosystems 5*(4), 187–196.

Von Neumann, J. and Burks, A. W. (1966). Theory of self-reproducing automata. *IEEE Transactions on Neural Networks 5*(1), 3–14.

Waldrop, M. (1994). *Complexity: The Emerging Science at the Edge of Order and Chaos*. London: Penguin Books.

Wells, J. (2013). *Complexity and Sustainability*. New York: Routledge.

5 Escaping Flatland

From a Newtonian to a complexity reality

Caminante, son tus huellas
el camino, y nada más;
caminante, no hay camino,
se hace camino al andar.
Al andar se hace camino,
y al volver la vista atrás
se ve la senda que nunca
se ha de pisar.
Caminante, no hay camino,
sino estelas en la mar.

[Wanderer, your footsteps are
the road, and nothing more;
wanderer, there is no road,
the road is made by walking.
By walking one makes the road,
and looking back
one sees the path
that will never be trod again.
Wanderer, there is no road,
only a wake upon the sea.]

(Antonio Machado, from 'Proverbios y cantares')

Confusing the map with the territory

Polish philosopher and scientist Alfred Korzybski is famed for his dictum "the map is not the territory" (1958), which refers to his observation that many people confuse the representation or abstraction of an object with the object itself. The theory of logical types (Whitehead and Russell 1913) demonstrates how system wholes are qualitatively (logically) different from their parts, and how paradox is generated when logical levels of phenomena are confused. Thus, when people confuse the map with the territory they are making an error in logical typing. In so doing they open the door for paradox (Bateson 2000), as we will see in the chapters that follow. Bateson (*ibid.*) applies a Kantian twist to Korzybski's allegory, arguing that it is impossible to truly know what territory, or reality, is:

> We say the map is different from the territory. But what is the territory? Operationally, somebody went out with a retina or a measuring stick and made representations which were then put on paper. What is on the paper map is a representation of what was in the retinal representation of the man who made the map; and as you push the question back, what you find is an infinite regress, an infinite series of maps. The territory never gets in at all … Always,

> the process of representation will filter it out so that the mental world is only maps of maps, ad infinitum.
>
> (Bateson 2000: 460–461)

Thus, all our descriptions of reality necessarily involve a level of abstraction and reduction (i.e. they cannot describe its entirety in full),[1] and therefore can never be more than 'maps'—we can never truly know reality. Attempting to invoke 'truth' as the goal of science (or religion, or any other ideology) is therefore a fruitless diversion (see Funtowicz and Ravetz 1993; Ravetz 2005), and the wild goose chase of the sustainability debate (see Meppem and Bourke 1999). The absurdity of attempting to do so is illustrated in Borges's *On Exactitude in Science* (written as a literary forgery):

> In that Empire, the Art of Cartography attained such Perfection that the map of a single Province occupied the entirety of a City, and the map of the Empire, the entirety of a Province. In time, those Unconscionable Maps no longer satisfied, and the Cartographers Guilds struck a Map of the Empire whose size was that of the Empire, and which coincided point for point with it. The following Generations, who were not so fond of the Study of Cartography as their Forebears had been, saw that that vast Map was Useless, and not without some Pitilessness was it, that they delivered it up to the Inclemencies of Sun and Winters. In the Deserts of the West, still today, there are Tattered Ruins of that Map, inhabited by Animals and Beggars; in all the Land there is no other Relic of the Disciplines of Geography.
>
> (Borges 1998: 325)

Nonetheless, 'maps' serve an important function: isolating certain signals from a 'territory' (i.e. reality), so we can navigate without being overwhelmed by its complexity. For example, in the case of actual (rather than metaphoric) maps, one might show topography, another might show roads. Neither are wrong; rather, both reflect different aspects of the same terrain and are more or less useful depending on how you are attempting to cross the territory (e.g. by car or by foot). Likewise Aboriginal Australians created verbal representations of territory using mythological metaphors in songs (Molnar and Meadows 2001), whereas Europeans did so spatially using paper maps—both approaches are useful for getting from A to B across the same territory. Aboriginal Australians also had a way of looking at the night sky that contrasted to that of Europeans, often naming the dark spaces between the stars (e.g. the Emu in the Sky) to present a different view of the same cosmos (Norris and Hamacher 2011). But perhaps the most illuminating example of the mind vs. map metaphor comes from science, in the quantum discovery that photons present as either waves or particles, depending on what the observer decides to measure. Thus, Heisenberg's statement "What we observe is not nature itself, but nature exposed to our method of questioning" (Heisenberg 1958: 81), recapitulates the notion that our observations can only ever be maps, abstracted representations, and *never* reality itself. This generates a dizzying circularity in our grasp of reality, which is perhaps best described by Gary Zukov, who stated:

> "Reality" is what we take to be true. What we take to be true is what we believe. What we believe is based upon our perceptions. What we perceive depends on what we look for. What we look for depends on what we think. What we think depends on what we perceive. What we perceive determines what we believe. What we believe determines what we take to be true. What we take to be true is our reality.
>
> (Zukov 2001: 344)

Maps are much more powerful than they seem on the surface because they do more than simply represent reality. Capra (1997), for example, describes how the habitual use of linear approximations to describe non-linear phenomena led many engineers and scientists to believe that decidedly non-linear natural phenomena were actually linear (see Chapter 4). In other words, confusing the map (linear approximations of non-linear phenomena) with the territory (non-linear phenomena) led to subconscious absorption of the assumptions underlying the map. So it happens that through repeated use, maps begin to shape how we perceive reality and therefore how we act into the future. Quantum physics and chaos theory demonstrate the profound interconnectivity of the universe and the vast, long-range influence even the most minute of actions can have. Because we cannot step outside the system, there is no objective observation. Thus, how we act based on our conceptions of reality, inevitably, and perhaps profoundly, affects the shape of future territory.

For example, Western industrial and post-industrial views of a linear natural world led to assumptions of unlimited energy and raw materials, and an infinite capacity on the part of the environment to absorb waste and pollution. This in turn led to the linear structuring of modern industrial food, energy and water and sanitation systems (Jones *et al.* 2010), These systems have radically changed the earth's biosphere and hence the 'territory' of our reality. When we confuse the map with the territory, we commit an error in logical typing. Consequently, paradoxes and vicious circles inevitably arise (Bateson 2000; Watzlawick *et al.* 2011), and hence our current global sustainability crisis. Thus, maps do not just represent reality, they also shape it through our participation. This is why powerful and vested interests attempt to dominate and control ideological discourse (as per Foucault 1980).

It is also why unearthing the subconscious maps that shape our assumptions about reality is so important when we face intractable problems. The importance of becoming aware of the assumptions that govern our behaviour is highlighted by an example from mathematical game theory showing that "if a person comes to 'know' a theory about his behaviour, he is no longer bound by it but becomes free to disobey it" (Howard 1971: xx). Without such awareness, our problem-solving abilities are caged by the paradigms within which we currently operate, as exemplified by the two-dimensional creatures inhabiting Abbott's (1884) satirical Flatland, who, even when confronted by an actual three-dimensional being (empirical proof!), cannot comprehend the notion of three-dimensional space. This is because their thinking is confined by an unassailable belief that the world is made up of only two dimensions, shored up by the collective weight of their past empirical observations. The wisdom of being sceptical of underlying assumptions was embodied within the brilliant mathematician Henri Poincaré, who, rather than working from pre-established principles as do most mathematicians, started from scratch in all his work (O'Connor and Robertson 2003). Which brings us to the second step in solving intractable problems: revising the deep assumptions that underpin our thinking about the problem situation. In this chapter, I will sum up our basic assumptions about the nature of knowledge and reality as shaped by centuries-old Newtonian thinking, and present an alternative set of premises and assumptions based on a synthesis of twentieth-century understandings from quantum theory, the complexity sciences and process philosophy.

The Newtonian paradigm

Newtonian thinking has done much to shape the Western 'map' of how the universe works. As outlined in Chapter 2, the Newtonian paradigm's dominance over the centuries has had the effect of subconsciously imbibing western minds with a 'watch maker' view of nature and a belief in a mechanical universe that acts like a complicated clockwork mechanism governed by linear mathematical laws (Capra 1997). Believing that the world consists of a collection of discrete things or entities governed by immutable laws has led to a number of fundamental assumptions which continue to influence how we perceive reality and knowledge. These assumptions were largely formulated in ancient Greek times, and then elaborated on during Europe's seventeenth-century scientific revolution. A summary of these assumptions, with their principal[2] origins indicated in brackets, is as follows:

- Reality consists of separate material things separated by void (fifth-century BC atomism, 17th century materialism).
- A separate external reality exists independently from ourselves; objective, neutral observation is possible (seventeenth-century materialism and empiricism).
- The mind is a material epiphenomenon of upward causation (seventeenth-century materialism), or a separate God-given non-material entity (seventeenth-century Cartesian dualism).
- The universe is governed by 'laws' that somehow pre-date its formation—fixed rules exist in the universe (third-century BC Plato, seventeenth-century classical mechanics).
- All phenomena are the consequence of upward causation, cause-and-effect; with complete knowledge we can predict and control events (fifth-century BC atomism, 17th century determinism and classical mechanics).
- The universe is deterministic; time is reversible, therefore all things can be disassembled and reassembled (seventeenth-century determinism and classical mechanics).

- Instability and change are caused by external disturbance (seventeenth-century classical mechanics).
- Phenomena are linear in nature; the whole is equal to the sum of the parts, so small disturbances create small changes, and big disturbances create big changes (seventeenth-century classical mechanics).
- All phenomena can be understood by deconstructing them into their component parts (seventeenth-century reductionism).

In a Newtonian universe, a system evolves as its elementary particles (atoms) move about, but there is no way for these particles to merge, disappear or divide. Everything (i.e. matter) that presently exists has always existed and will always exist, albeit in different configurations (Heylighen *et al.* 2007), moving around inside an inert container of space and time in accordance with universal rules and laws that pre-date the universe. As such, there is no true novelty or creation (Prigogine and Stengers 1984), just rearrangement of matter through cause and effect. Any given state of the system is the inevitable cause-and-effect result of conditions in its prior state, and is therefore predicable given sufficient information. In other words, all phenomena, including ourselves and our minds, are equal to the sum of their material parts. Chaos, creativity, holism and the flow of reality have no place in this worldview.

Early-twentieth-century advances in relativity, quantum mechanics and systems theory (as described in Chapter 2), however, exploded many of the Newtonian paradigm's core assumptions, lending greater credence to the immaterial and subjective nature of reality posited by idealist and panpsychic philosophies. It also saw Whitehead's *Process and Reality*, one of the most influential philosophical works of the twentieth century, devoted to the shift from 'being' toward 'becoming'.[3] This was followed by a number of pivotal scientific advances in the latter half of the century, which are now becoming integrated under the banner of complexity sciences, including chaos in non-linear dynamical systems and a scientific foundation for holism and emergence via complexity theory and other theories of self-organisation (Heylighen *et al.* 2007). Yet despite these developments, the Newtonian paradigm continues to dominate Western thinking. With its comparative simplicity and intuitive appeal, it has been so pervasive that most people today still equate scientific thinking with Newtonian thinking (*ibid.*).

Drawing a new map: a complexity-based ontology

The history of Western science and philosophy can be viewed as an ideological competition to control metaphysical discourse over the 'true' nature of reality, one that has culminated in the present-day domination of the Newtonian perspective (Chapter 2). Since the dawn of the scientific revolution, this competition has created debate between materialists and idealists, the empiricists and rationalists, Kantian philosophers and the logical positivists, and, more recently, between modern positivists and postmodernists. The quantum physicist Rovelli (2003: 28) attributes this "tortuous historical evolution" to modern science's "curious schizophrenic attitude of being anti-realist with electrons and iron realist with chairs".

Ilya Prigogine states that the astonishing success of science has led to a "collision between what has often been called the 'two cultures,' science and the humanities", and has led to the repeated question, "How to choose?" (Prigogine and Stengers 1984: 11). He also notes, however, that according to Alfred Whitehead, a clash of doctrines is an opportunity, not a disaster (Whitehead 1948). Rose (2003) states that for any living phenomenon there are many possible explanations or descriptions. So perhaps the opportunity lies in realising we don't have to choose, that it's not a binary decision between whose depiction of reality is true or untrue. Galois's mathematical group theory shows how opposites combine to reinforce the identity or pattern of the group to which they both belong. This phenomenon is likewise an element of Heraclitus's unity of opposites, and a feature of Hegel's dialectical reasoning, whereby thesis and antithesis are partial elements of a whole truth. It is also echoed in the quantum notion that matter and energy come into being through a unity of opposing properties. These observations, common to math, science and philosophy, hint that the opposing themes that repeatedly cycle through Western metaphysics may not actually be in conflict. Rather they are each, perhaps, aspects of a singular, higher-order phenomenon—glimpses of reality's paradoxical 'Janus face' illuminated from different angles. In this sense, the material Newtonian ontology and the immaterial idealist/panpsychic ontology are two sides of the same coin. They, along with other paradigms, are different 'maps' attempting to describe the same 'territory'.

The previous chapters reveal the convergence of a number of analogous themes, ideas and theories from mathematical logic, physics, the complexity sciences and process philosophy. Together, they present an opportunity to unify the objective material reductionism of the Newtonian ontology with the subjective immaterial holism of idealist and pan-psychic ontologies. They do so by providing explanations for how reality can simultaneously be material and immaterial, deterministic and deeply uncertain, and both dependent and independent of our minds and selves. In the following sections I outline the features of a proposed complexity-based ontology and epistemology, synthesised from complementary concepts introduced in the previous chapters, namely: mathematical logic, quantum theory, chaos and complexity theory, autopoiesis and process philosophy. Together this ontology and epistemology comprise a coherent complexity-based paradigm,[4] one which serves as a 'more complete'[5] alternative to the 'lop-sided' Newtonian paradigm by looking towards the 'whole' as well as the 'parts'.

Complex reality

This proposed complexity-based ontology is founded on Wells's (2013) argument that complexity is a profound characteristic of reality and hence of *all* systems. This differs from the Newtonian perspective, in which most natural phenomena are understood to behave in linear ways. Henceforth, all 'entities', 'phenomena' and 'systems' referred to in the text that follows are to be understood as *complex* (as per Chapter 4).

Immaterial processes

In a complexity-based ontology reality is seen to consist of immaterial interconnections, and is better understood as *processes* rather than things. As postulated by quantum theory, things we once believed to be material, such as atoms, are actually just interconnecting processes between things, which themselves consist of interconnecting processes between other things, and so on (Bohr 1958). This pertains to Bohm's (2012) notion of implicate order in quantum theory. He describes how it is the relationship between things and not the things themselves that give rise to material reality. Objects appear to us as solid and material by virtue of scale, but are revealed as immaterial patterns of relationship when examined at the quantum level. Because of the scale at which our human awareness operates, we cannot 'see' the insubstantial nature of matter with our senses. Time also creates the illusion of a substantial reality—stable patterns of relationship that persist in time also appear to us as substantial entities (Rescher 2008). This quantum perspective is in keeping with autopoiesis (Maturana and Varela 1980) and process philosophy (Whitehead and Sherburne 1981), which likewise postulate that reality has no material foundation. What appear to be substantial are actually processes organising into spatial and temporal patterns—all entities are just units of interaction.

Correspondingly, from this point on, when the terms 'entity' or 'system' are used in this book they will connote an emerging coherent pattern of immaterial *relationships*, rather than what they normally signify, which is a configuration or network of substantial *things.* Likewise, when the term *agent* it used, it will be understood as a *sub-pattern* of coherent relationships that is subordinate to a higher-order unifying pattern. This approach is consistent with Checkland *et al.*'s (1981) soft systems methodology, which treats systems as epistemological entities (i.e. symbolic mental representations of something experienced in reality, or 'maps'), rather than as ontological entities (i.e. actual things, or 'territory').

Interconnected whole

Quantum theory has shown that when viewed at atomic and subatomic scales, reality is in fact a "complicated web of relations between various parts of a unified whole" (Capra 1997: 30). Heisenberg (1958: 107) similarly states "the world thus appears as a complicated tissue of events, in which connections of different kinds alternate or overlap or combine and thereby determine the texture of the whole". Bohm (2012) argues that this network of relationships is an underlying *implicate* order that is enfolded in the whole of the universe. *Explicate* order, on the other hand, consists of those 'things' we can observe and measure, and is generated by implicate order, through the relationships between things. Bohm believes that quantum discoveries have brought about insights that "can perhaps best be called Undivided Wholeness in Flowing Movement" (*ibid.*: 11). His view implies that flow is in some sense prior to that of the 'things' that can be seen to form and dissolve in this flow. He likens this to vortices in a stream—a vortex is a stable pattern within a continuous flow; the vortex is undivided from and contiguous with the stream, which itself (as the whole) is in a constant state of flux. These notions of quantum wholeness mirror the central thread of interconnectedness and flux that weaves through Heraclitus's philosophy (you can never step into the same river twice), and Whitehead's assertion that the physical world is bound together into a continuum "by a general type of relatedness" (Whitehead and Sherburne 1981: 81).

Observer dependent

In the Newtonian paradigm, a separate external reality exists independently from ourselves, allowing objective and neutral observation of that reality. At the quantum scale, however, atoms and particles are not isolated 'things', but rather "abstractions, their properties being definable and observable *only through their interactions with other systems*" (Bohr 1958: 57; italics added). Quantum physics demonstrates that observing a system necessarily changes a system. Hence, at the quantum scale, the idea of a scientist playing the role of detached observer is shattered, along with the Newtonian assumption of an objective reality. By setting up what it is we are going to measure and observe, we determine to some extent the properties of the observed system: "The universe will never afterwards be the same. To describe what has happened, one has to cross out that old word 'observer' and put in its place the new word 'participator'. In some strange sense the universe is a participatory universe" (Wheeler 1973: 244).

The vast interconnected wholeness of reality implied by quantum theory is also a feature of chaos theory. Chaos theory demonstrates the extreme sensitivity dynamic non-linear systems have to initial conditions, due to bifurcations caused by amplification of small events. As such, the scientific ideal whereby an observer occupies a privileged, objective position outside a system is impossible. If both reality and our experience of it are necessarily shaped by how we observe it, notions of subjective versus objective reality are meaningless. Ultimately there is no outside or inside because all entities are processes interconnected within the tissue of reality. Indeed, there are no truly closed systems other than the entirety of the universe.

Indeterminate and irreversible

In the Newtonian paradigm, reality is deterministic, the inevitable result of cause-and-effect (upward causality), unfolding in accordance with immutable laws that somehow predate the universe's existence. In this view, the arrow of time and hence all processes are theoretically reversible. Once the first kernel of the universe is set into motion, true novelty is no longer possible. With sufficient knowledge about the state of a system, we can accurately predict its future states, and back-cast to know its past states as well (as per Laplace 1951: 4). The inexorable disordering of closed systems via entropy in classical thermodynamics, and the complex irreducible ordering of open systems via bifurcations in non-linear thermodynamics, however, suggest the arrow of time is *irreversible* (Prigogine and Stengers 1984).[6] According to Prigogine, chaotic bifurcations are points in a complex system's evolution where many futures are suddenly possible and change becomes *a causal*. This makes it impossible to predict with certainty which direction the system will take next (Prigogine 1997). The period prior to wave function collapse in quantum systems is likewise a point where many system states are possible. Again, it is impossible to predict with certainty which state will emerge from the collapse. When combined, quantum and chaotic uncertainty create a potent indeterminacy:

> It is the exponential amplification of errors due to chaotic dynamics that provides the second reason for Laplace's undoing. Quantum mechanics implies that initial measurements are always uncertain, and chaos ensures that the uncertainties will quickly overwhelm the ability to make predictions. Without chaos Laplace might have hoped that errors would remain bounded, or at least grow slowly enough to allow him to make predictions over a long period. With chaos, predictions are rapidly doomed to gross inaccuracy.
>
> (Crutchfield *et al.* 1986: 50)

Prigogine believes that by clinging to cosmic reversibility, quantum and classical physicists are idealising nature. His arguments (as recounted by Briggs and Peat 1989) are that all natural systems are indeterminate, and that the laws of the universe are not given, or static and unchanging. Rather they emerge and evolve as bifurcations and amplifications increase systems' complexity through time, a view that has raised the ire of some physicists (see Pagels 1985). For Prigogine, the irreversibility that emerges from system instabilities may very well be the source of order and coherence in the universe.

Change versus creation

From the perspective of a complexity-based ontology, reality is continuously creating itself—dynamically emerging from the creative interactions of the entities that comprise it. In other words, the universe is being *co-created* at every moment through a process of action and reaction. Entities (which themselves are just nested patterns of interaction) perpetually interact to co-evolve and co-create reality, bringing forth the future, as per Whitehead and Sherburne's (1981) process philosophy, Maturana and Varela's (1980) autopoiesis and Kauffman's (1993) fitness landscapes. As such, each new moment is novel, inevitably different from the previous. This resonates with Heraclitus's notion of the ever-changing flow of reality, and his dictum that "you can never step in the same river twice". It also relates back to quantum theory's observer dependence and Wheeler's (1973) notion of a participatory universe. We participate in the universe's co-creation at every moment through the process of our relating, the effects of which ripple through the web of existence.

This view of reality and the universe as being perpetually constructed, or *becoming*, differs subtly but significantly from the Newtonian perspective. In the Newtonian view, the universe is a shifting configuration of matter and energy set against the backdrop of absolute space and time. Change is an alteration in some aspect of the universe's pre-existing configuration, the result of disturbance from some outside force. In the proposed complexity-based ontology, however, what we perceive as change is actually *creation*—creation of a new moment that differs from the last. Grasping reality as *emerging* rather than changing requires a logical jump in how we think about time. Instead of thinking about time and space as separate entities (as per Newtonian classical mechanics), one must think of them as being unified in the fourth dimension as spacetime (as per Einstein's special relativity). In other words, it requires that we see time as integrated with space into a *flow of events*, rather than as a separate unrelated dimension. Every present moment is a configuration of related events that are the manifestation of a 'growing' reality. (Although I will continue to use the word *organisation*—the act of arranging, i.e. changing, something—in this context the term is better understood as *creation*, and self-organisation better understood as *self-creation*.)

This is difficult, however, because our grip on the dimension of time is tenuous at best—we have only the highly limited and unreliable recall of our diaphanous memories. As a result, we cannot at once hold in our minds the continuous flow (i.e. pattern) of events that precede the present moment. One way of gaining a grasp of time as a flow of events is to examine stroboscopic photographs showing patterns of motion. Figure 5.1a shows a stroboscopic image of a bouncing ball. In a normal photo, a bounced ball would appear as a circle suspended in space, and would appear to have only two spatial dimensions. In the stroboscopic photo, however, the dimension of time is captured along with that of 2D space, and thus the pattern of the moving ball appears as a more complex three-dimensional spiral[7] (this image also helps us understand how a quantum particle might appear to behave as either a wave or particle, depending on how we observe it). By capturing a flow of events in sequential freeze-frame images, the stroboscopic effect creates a perceptual jump to a higher logical order. This perceptual jump can help us understand space and time as being unified higher-order phenomena, rather than as separate and unrelated lower-order phenomena.

Similarly, by capturing time as well as space the stroboscopic sculpture of a jumper in motion in Figure 5.1b creates the impression of the entity *growing into* a three-dimensional space, or emerging. Our normal perception would be that of the jumper *moving around* in space, as opposed to growing into it.[8] The scale at which we experience time causes us to perceive most things as existing and moving around in space, rather than emerging through self-creation. If we lived much longer, and our perception of time was radically sped up, before our eyes we would see all things—including animals, trees, forests, mountains, and planets—emerging and growing, then breaking down and disintegrating before rising up once again. We could then comprehend reality as continuously emerging in a dynamic process of creative destruction, of oscillating order and disorder.[9]

If one visualises 'flattening' a perpetually emerging and self-creating universe (by removing the fourth dimension of *unified spacetime* and thereby dropping

(a)

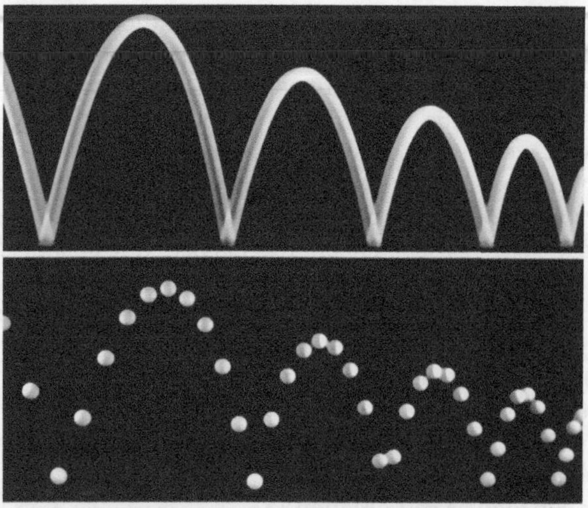

(b)

Figure 5.1 (a) Stroboscopic image of a bouncing ball, illustrating how a particle can be understood as either a particle or a wave. (b) Stroboscopic sculpture illustrating a shift in perception from that of an object changing position, to that of an object growing into a space. By capturing the dimension of time with that of space, these images help us make the perceptual 'jump' to a higher logical order whereby we can recognise space and time as unified rather than separate phenomena

Source: (a) photo by Prof. Andrew Davidhazy (retired), School of Arts and Sciences, Rochester Institute of Technology; (b): 'The Long Jump', scuplture by Peter Jansen (www.humanmotions.com).

down a logical level), the result is a Newtonian universe, where space and time are separate, unrelated entities, operating within the same dimension (as per Figure 2.1a). As such, our adherence to the Newtonian view equates to using a two-dimensional map to describe a three-dimensional territory, then coming to believe that the map *is* the territory, that the territory is also two-dimensional. This is a case of confusing logical types—assuming unified spacetime is of the same logical level as space. As discussed in Chapter 3, the consequence is paradox, such as the paradox of instantaneous gravitational force.

Emergent spacetime

As noted above, in the Newtonian perspective space and time are absolute. They predate the universe and serve as a container for it. However, from the complexity-based understanding of a perpetually *emerging* universe, space and time no longer exist in an absolute sense. Unified spacetime is *created* by the dynamic interactions of an ever-emerging reality, rather than being a container for it. Or, more correctly, spacetime is a mental construct used to describe the *scope* of this emerging reality[10]—it is not an absolute thing of itself. This relates to quantum gravity theories, which propose that four-dimensional spacetime is a higher-order phenomenon that somehow emerges from deeper reality.

One way of understanding this differentiation is by using billiards as a metaphor for visualising a *changing* universe from the perspective of Newtonian ontology. The billiard table is a two-dimensional representation of the three-dimensional space and time backdrop of the Newtonian universe. Time is visualised as operating separately but in the same dimension as space (i.e. the plane of the billiard table), as per Figure 2.1a. The player striking the cue ball represents the mysterious first force that sets universe in motion (i.e. the Big Bang). The billiard balls, when racked, are the equivalent of the primordial unity that, once struck, explodes into atomic bits of matter. As the balls break apart, they represent lumps of matter colliding around in space and time until gradually slowed to a complete stop by friction/entropy.

In contrast, a useful metaphor for visualising an emergent *self-creating* universe is the inflating cascade of bubbles created when soap is poured into fast-flowing water. The bubbles of surface tension created as polarised soap molecules spontaneously align are equivalent to higher-order patterns emerging from creative interactions between quantum entities. The growing three-dimensional bubble mass can be understood as a lower-order representation of emerging four-dimensional spacetime, or reality. The spatial scope of the universe is represented by the breadth of the bubble mass. Its temporal scope is represented by the bubble mass height. As the bubble mass spreads horizontally in the second dimension, it represents emerging space. As it rises vertically in the third dimension it represents emerging time. Together they represent emerging events (space + time), as per Figure 2.1b. Thus, we can see how space and time are simply linguistic and mathematical constructs (i.e. maps) we use to describe the *emerging shape of reality*.

United in the fourth dimension

Another important aspect of spacetime is its *unifying* property, as per Figure 2.1. Our intuition of a unifying four-dimensional spacetime can be developed with the following exercise. Examine the images in Figure 5.2. Image (a) depicts five objects that appear to have no connection—they are separate. Image (b) depicts two objects that likewise appear to have no connection. Nor is there any apparent connection between image (a) and image (b).

However, the objects within each image and the two images themselves *are in fact connected*. It is impossible, however, to discern this connection from the images presented. This is because they are two-dimensional viewpoints of a three-dimensional phenomenon—they are of a lower logical order than the whole phenomenon of which they are a part. The three-dimensional viewpoint is illustrated in Figure 5.3.

In Figure 5.3, the apparently separate objects depicted in Figure 5.2 are revealed as connected in the third dimension. This is by way of being two-dimensional transverse sections[11] (parts) of a three-dimensional human. Using the logic of Bateson (2000; see Chapter 3), because a transverse section is of a lower-order dimension (x, y) than the whole human (x, y, z), a whole human cannot be inferred or understood from just looking at a transverse section of a human.[12] Nor can it be inferred by simply looking at more transverse sections of human. For example, if, as an observer, one were to:

1 move *through* the whole human along the z-axis, from top to bottom, but from *inside* the two-dimensional (x, y) perspective, our experience would be one of watching abstract objects *moving around and reconfiguring* in space.[13]

We do not get a meaningful understanding of the whole human until we step out from inside the body, and view the whole human from an *outside* three-dimensional perspective. Thus, if an observer were to:

2 stand in the third dimension *outside* the transverse sections and observe them stacking up from feet to head (i.e. introducing *height*, sequentially from the bottom to the top of the z-axis arrow in Figure 5.3) our experience would be one of watching a human *emerge* from the ground up.

Understanding the whole human requires a *jump* to a higher dimension/logical order—that of the third dimension (x, y, z). Note that the x, y and z dimensions are not separate unrelated components—they are nested and continuously interrelated aspects of the whole. The logic from this exercise can then be applied to how we think about space and time versus spacetime, and therefore the universe:

z (height)

2D transverse section (x, y) = *part* of a human ⟶ 3D body (x, y, z) = *whole* human

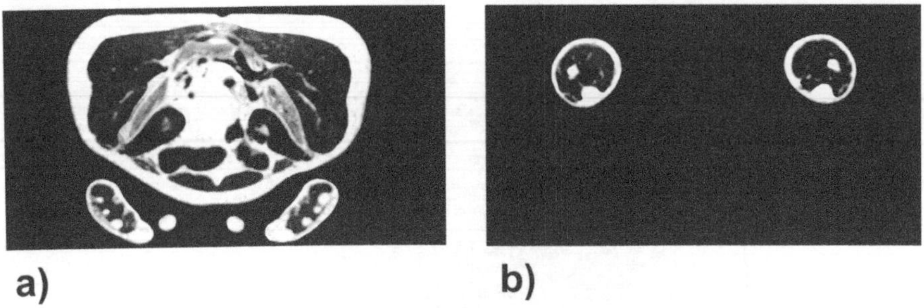

Figure 5.2 Two images depicting objects with no apparent connections
Source: adapted from Mad Sci Network (2013).

Thus,

$$\text{3D space } (x, y, z) = \textit{part} \text{ of the universe} \xrightarrow{t \text{ (time)}} \text{4D spacetime } (x, y, z, t) = \textit{whole} \text{ universe}$$

This brings home the notion that four-dimensional spacetime is not a *component* of the universe—it *is* the universe: the whole dynamically emerging universe, from its origin to the present moment. Thus, observer experience (1) outlined above would equate to being *inside* the universe and time-travelling *through* it, from its beginning to the present. Observer experience (2) would equate to standing *outside* the universe and watching it emerge/self-create, inflating into four dimensions like a cascading bubble mass.[14]

Our attempts, therefore, to understand the universe from a Newtonian perspective (which is our everyday experience of it) are equivalent to trying to understand a whole human from what we can see in a transverse section of it (Figure 5.2). The only way of gaining a meaningful understanding of the *whole* universe is to step *outside* of it, to view it from a unified four-dimensional perspective. This act is beyond the capacity of our senses, given the logical level at which humans operate in the cosmos.

Irreducible: the whole is greater than the sum of its parts

Understanding system wholes also requires an understanding of system *irreducibility*. Even though no moment is a perfect replica of the previous, stable probabilistic patterns do emerge in reality when different entities/systems couple together to create feedback loops. In doing so, they harmonise their interactions and self-organise (i.e. self-*create*) to generate a higher-order pattern of behaviour that unifies the system into a whole. The whole is greater than the sum of its parts,[15] as per holism, general system theory, quantum theory, Gödel's incompleteness theorem, the theory of logical types and complexity theory. As such its pattern is also qualitatively different to the patterns governing its subsystems (Richardson

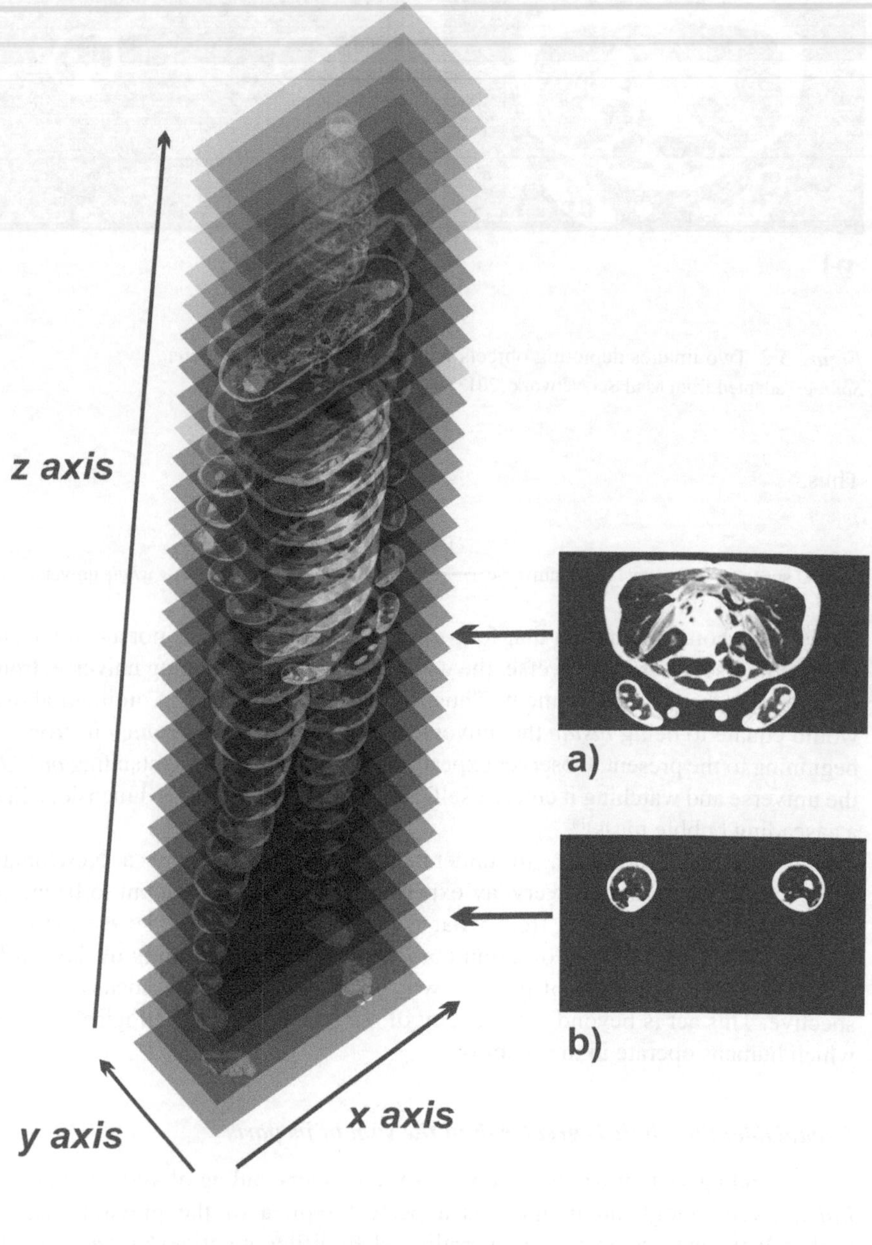

Figure 5.3 Two-dimensional transverse sections of a human body stacked to reveal a three-dimensional whole human, revealing how the objects shown in images (a) and (b) are connected

Source: adapted from Mad Sci Network (2013).

2005)—the pattern of a human is of a higher order and qualitatively different to the patterns of its organs and cells. Heylighen *et al.* (2007) use the example of sodium chloride (NaCl)—table salt. The essential qualities which define salt—salty, inert, edible and crystalline—are *emergent* properties, which are qualitatively different from the properties of its component parts: sodium, a volatile soft metal, and chlorine, a poisonous gas. It isn't the 'things' sodium and chloride that give rise to the 'thing' salt—it is the relationship between them, Thus, the whole always possesses a property or pattern that is not present in its parts, as per the role of missing information in Gödel's incompleteness theorem. As a consequence, because the system whole is greater than the sum of its parts, *coupling allows it to explore a greater phase space than would be available to the sum of its component subsystems*. This provides further evidence that the arrow of time is irreversible.

Dynamic nested patterns of relationship

When systems couple with other systems they create a metasystem, from whose interactions a governing metapattern spontaneously emerges (via upward causation). In turn, this metapattern exerts downward influence (downward causation) on the nested systems that comprise it (Heylighen *et al.* 2007). Hence, systems coupled into metasystems are constrained by the downward causation exerted by the whole. This means they have to obey the rules of relationship that maintain the coherence (i.e. emergent properties or circular organisation) of the metasystem, and can no longer act independently (*ibid.*). The cycle repeats itself when metasystems couple together to create meta-metasystems, and even higher-order patterns of circular governance. Hence, reality perpetually emerges as a dynamic pattern of nested relationships.

Richardson (2005) notes that each level or dimension of organisation "exhibits a substantial realism" (*ibid.*: 623), meaning there are "solid representations of aspects of complex systems that need not include the whole system" (*ibid.*: 624). He lightly suggests this could be an argument for a kind of "soft reductionism". Indeed, because of their interconnectivity, descriptions of complex systems inevitably require a level of reduction, as it is practicably impossible to describe the whole system[16] (i.e. the universe; Richardson 2005).

Upward + downward causality = circular organisation

This dynamic combination of upward and downward causation is equivalent to the autopoietic notion of *circular organisation* (Maturana and Varela 1980)—the generator of homeostasis and the hallmark of living systems. Circular organisation can only be established and maintained in open systems where there is a continuous through-flow of material and/or energy (i.e. systems far from equilibrium, as per Prigogine and Stengers 1984). Once established, circular organisation serves to maintain the subsystem interactions that created it (via negative feedback). This involves reinforcing and reproducing the 'identity' or cohesion of the system whole by exerting a higher-order governing pattern (downward causality) onto its component subsystems such that their interactions continue to reproduce the higher-order circular organisation (upward causality). This reflects Heraclitus's unity of opposites dictum that the 'the path up and the path down are one in same' (Capra 1997).

As such, the system is self-referential, or self-reproducing. With repeated iterations it tends to become increasingly stable and cohesive over time. This is why it is often difficult to change emergent order once it has established. A persistent pattern of relationship creates the sense of a *boundary* (Richardson 2005), giving it the impression of substance and separateness from the perspective of the observer. Boundaries are thus emergent, temporary, and relative to the scale of observation (*ibid.*).

Circular organisation is equivalent to the system's attractor. It is the subset of phase space to which the system is attracted as a result of coupling and feedback between its component subsystems. The interactions of the subsystems form patterns, which become rules or—more accurately—*habits* governing their interactions at the subsystem level. These patterns of interaction exert upward causation to form the system's metapattern (note that the system *is* the metapattern, the metapattern is not just a property of a system), which then exerts downward causation onto the component subsystems/subpatterns. Together they create a circular governing pattern of the system whole, one which is *qualitatively* different from the patterns governing its subsystems.

Concurrent persistence and change

Thus, even though its subsystems continuously interact and change, the overall circularity (i.e. identity) of the higher-order system is maintained. This reflects Galois's group theory, which demonstrates how marked changes taking place *within* a group do not cause any change in the *overall* group itself (see Chapter 3). Indeed, the more entities interact and change at the subsystem level, the more they reinforce the circular governing pattern of the system whole, 'feeding' the cybernetic feedback loops that maintain system homeostasis. This results in the paradoxical coexistence of persistence and change[17] illustrated in the French proverb "the more something changes the more it stays the same" (Watzlawick *et al.* 2011). It also pertains to Heraclitus's unity of opposites: despite their apparent opposition, persistence and change must occur together, as one cannot exist without the other. Watzlawick *et al.* (*ibid.*) illustrate by noting that if everything in the universe were blue, it would not be possible to have a concept of 'blueness'. Phenomena only become apparent (i.e. distinguished) through contrast and comparison.

History constrains rather than determines the future

Because circular organisation is self-reinforcing, each new moment is *likely* (but not certain) to be *similar* to the previous one. The shape of each moment constrains the possibilities (i.e. phase space) of the next,[18] as per process philosophy (Whitehead and Sherburne 1981) and chaos theory. Paradoxically, however, within these constraints (which form the shape of the system's attractor in phase space), an infinite number of possible states are still available to the system due to sensitivity to initial conditions and spontaneous emergence. Thus, while each new moment *tends* to be similar to the previous, it will *never* be exactly the same, meaning it is fractal in nature. In the Newtonian ontology, historic events *determine* future events. In a complexity-based ontology history *constrains* rather than *determines* the future.

Order and chaos are co-extant

Apparently stable patterns are not immutable, however, as they are perpetually co-evolving with other entities. As discussed in Chapter 4, complex systems can also become radically unpredictable in certain situations and spontaneously undergo dramatic transformation. This is because stability and instability co-exist in complex systems: yet another reflection of Heraclitus's unity of opposites. Stable patterns spontaneously emerge in chaotic systems. Likewise, given a sufficient number of iterations, stable patterns eventually and spontaneously become chaotic, even in apparently determinate systems. This is particularly the case when (i) rates of energy, material or information flowing through the system continue to increase, amplifying internal instabilities, and/or when (ii) the system increases its internal and/or external connections/couplings, creating new feedback relationships that disrupt the system's existing circular pattern. In these situations, it becomes more difficult for the system's circular organisation to apply sufficient negative feedback to damp down disturbances. This makes the system susceptible to destabilising amplification due to positive feedback.

System transformation and the edge of chaos

As a system becomes destabilised it begins to approach the *edge of chaos*. At this point, a small disturbance can be amplified to the point where the system's behaviour quickly becomes chaotic, or turbulent. The increasingly chaotic state of the system indicates a *transition or bifurcation point*, where the system starts to oscillate between its original governing pattern and new potential governing patterns. New governing patterns begin emerging as a result of increases in number and types of couplings, either between the system's component subsystems, or between the system whole and other systems. If its control parameters continue to 'dial up' (i.e. their rates increase), the system begins to wildly oscillate between other dimensions of order (i.e. basins of attraction, or attractors) potentially available to it (Briggs and Peat 1989). In phase space, the surface of the system's attractor (i.e. circular organisation) breaks up as it enters the fractal dimension of indecision and begins to explore the wider phase space available to it (*ibid.*). At this point the system can potentially move in three general directions:

1 *Loss of order:* The system is unable to harmonise/self-organise its new interactions/couplings into a new level of order before the original coupling

patterns that maintain its present circular organisation are irreversibly destroyed. This causes the system to break apart into its component subsystems through total loss of its circular organisation/attractor.

2 *Maintenance of existing order:* The original circular organisation/attractor manages to dampen out the effects of any new coupling/interactions, causing the system to collapse into its original attractor/pattern. This is unlikely unless the system's control parameters are 'dialled down' in time.

3 *Increase or change in order:* the new couplings between the interacting subsystems and/or systems will spontaneously harmonise and create a completely novel (qualitatively different) higher-order governing pattern or circular organisation (i.e. a strange attractor).

Scale dependence of chaos and order

Both quantitative and qualitative properties of a complex systems are scale dependent. This is illustrated by Mandelbrot's (1967) question: "How long is the coast of Britain?" Because the coast is jagged and this jaggedness is self-similar at different scales (i.e. it is fractal), it depends on the size of the units you use to measure it. The coastline is indefinitely long if you use an indefinitely small unit to measure it (see Chapter 4); hence, quantities are dependent on the scale of measurement. Jaggedness is a qualitative measure of the system's chaos and is also relative to scale. For example, if you look very closely at the water's edge it appears jagged. If you stand in a bay and look down the coast the shoreline looks relatively smooth. If you view the coastline from a low-flying plane it appears jagged. If you look at it from higher in space the coastline once again looks smooth. Thus chaos and order are laminated and scale dependent. This also works on a temporal scale. Events viewed in a short timeframe may appear random or chaotic. When viewed at longer timescales, however, they are often revealed as part of a larger pattern of order that is not apparent in shorter time periods.

As such, the apparent stability or instability of a system is a function of the scale of observation. Imagine yourself observing the dynamic webs of nested systems that comprise reality (system = group of coupled subsystems) from the vantage point of being inside the web, but with you growing bigger (like Alice in Wonderland). The system appears stable when you are viewing it at a spatial or temporal scale whereby the patterns governing it are obvious, i.e. when you are looking at just *one* level or dimension of the system at a time—looking at a tree, for example. However, as you 'grow' and begin to move into a higher spatial scale, the system starts to appear unstable/chaotic as more and more trees come into your field of vision, in no discernible order.[19] Now you are observing at the scale of the transition point *between* system levels/dimensions. You are seeing a 'mash-up' of qualitatively different governing patterns from the two different system levels—you can't see the forest for the trees or the trees for the forest. At this point your perception of the system is shifting from one dimension to another, or from one level of organisation (or logical class) to another, creating the sensation of chaos or turbulence. The system appears to 'flip-flop' between the two dimensions as you discern the governing pattern from one level, then the other. The governing patterns of both the lower and the higher dimension/level of organisation are not simultaneously apparent to the observer. This is because being qualitatively different they have to be measured/observed with different 'tools'. This premise is reflected in the theory of logical types, the phenomenon of emergence in complexity theory, and the principle of complementarity in quantum theory.

As you keep 'growing' through the system, your field of vision eventually extends out and above the trees, allowing you to see the next system level—the forest. You can no longer discern the apparent chaos of the transition from the lower system, nor can you discern individual trees. Again, as per the theory of logical types and emergence, the higher-order pattern is qualitatively different from the lower, and cannot be measured or understood using the measures and terms used to describe the patterns at the previous, lower level. For example, you cannot describe the pattern of a forest (e.g. area, connectivity, diversity, composition, edge, dispersion) using the descriptors (qualitative or quantitative) used to describe the pattern of a tree (e.g. height, circumference, trunk, branches, stems, roots, leaves, flowers). In addition, the forest's pattern is not only influenced by the interactions between the trees and other biophysical entities comprising it, but also by other landscape level systems such as adjacent vegetation communities, fire, and topography. These systems interact with the forest to create larger metapatterns across the landscape, which in turn exert larger governing patterns onto the lower level composite systems. These metapatterns come into view as the observer continues to 'step back' from the system to view it from an increasingly large scale.

Larger-still governing patterns come into play at the biosphere level, including atmospheric, hydrological and geological forces, as per Gaia theory (Lovelock and Margulis 1974). At every level, the patterns that emerge are the result of the co-evolutionary dynamic interplay between higher and lower-order systems/patterns. These patterns represent higher unified dimensions of order, as conceptually illustrated in Figures 2.1 and 5.3. What appear to be isolated[20] or disordered phenomena at one spatial or temporal scale are shown to be unified into larger system patterns at another scale. A system appears isolated to an observer when they cannot see the larger governing pattern that unifies or controls the macro-behaviour of the system. Likewise, a system appears disordered when viewed at a *spatial* transition point (as described above), where the observer sees a confusing jumble of governing patterns from different levels/dimensions. When viewed at a *temporal* transition point, where it is poised on the edge of chaos, a system also appears disordered. This is because it is undecided and 'jumping' between new possible states before transforming into a higher-order pattern.

Scale dependence of rules and laws

Thus, from this proposed complexity perspective, seemingly immutable laws governing the universe are actually just patterns given the illusion of immutability by virtue of their persistence in spacetime relative to the observer. This is reflected in Chew's 'bootstrap hypothesis'[21] of the universe (Chew 1968), which rejects the existence of any fundamental properties or entities, including laws, constants, and equations. In the bootstrap view, these properties are understood as flowing from the dynamic web of interconnected events that comprise the universe (Capra 2010)—that is, they are emergent.

Likewise, system boundaries are also scale dependent. To the naked eye, the skin appears to be a substantial and impermeable boundary separating us from our environment. However, viewed at a nano-scale skin is semipermeable, allowing exchange of fat-soluble substances, water and small nano-particles. Meanwhile, at the quantum scale, skin is revealed as an immaterial pattern of relationships interconnected with the totality of the universe. Thus all seemingly unrelated phenomena are interconnected at some level, bringing us back to the idea that there can be no objective observation, that there is no external reality, only a vast tissue of nested interconnections of which we are a part. Our every thought and action reverberate through this web, however indiscernibly, irrevocably changing the emerging pattern of reality, which we understand to be the universe.

Scale dependence also relates to quantum physics, in that the universe's 'rules' of operation change when you start observing things at a very small scale (Prigogine and Stengers 1984). In essence, the observer jumps down a dimension/logical level to view the system at a quantum scale. At the quantum scale, governing patterns of macro-reality's subsystems can now be discerned. As per the theory of logical types, the governing patterns at the quantum level are of a qualitatively different nature to the metapattern operating at the level of classical Newtonian mechanics. Penrose (1995) notes that quantum physics has not rendered classical physics wrong—it simply operates at a different level. Quantum physics cannot be used to explain the classical level of reality any more than classical physics can describe the quantum level. Rather, it's a matter of recognising that phenomena behave very differently at different scales. For example, it would be impossible to recognise the organisational patterns of a human system at the quantum level, or to explain them using the rules and premises that comprise quantum theory. As Heisenberg (1958: 81) stated, "What we observe is not nature itself, but nature exposed to our method of questioning".

Evolution through self-organisation

A complexity-based ontology also explains why living systems become more complex over time, and overturns the traditional understanding of natural selection as the main driver of evolution. Isolated (closed) systems progress towards increasing disorder as per the second law of thermodynamics, In contrast, open systems in continuous interaction with their surroundings progress towards *increasing* and often transformational order, as per non-linear thermodynamics and complexity theory (Chapter 4). These open systems not only become transformed, they also transform and are transformed by the other systems that comprise their 'environments', each jockeying to maintain their own circular organisations or identities as they interact.

As such, self-organising emergent order ('order for free') generates novelty and drives living systems to become more complex over time. According to Kauffman

(1993), natural selection works to limit novelty by *squashing* new emerging order, rather than creating it. As systems become more complex through successive emergent transformations, they become more resistant to disturbance and selective pressures (*ibid.*). In this sense, natural selection can be understood as the failed harmonisation of two systems, where one system maintains its circular organisation (identity) at the expense of another.

Lovelock and Margulis's Gaia hypothesis (1974) also counters the classical Darwinian premise that organisms evolve as they adapt to their environment. This idea implies that an organism's environment is a somewhat static backdrop which exists separately and externally to it. But in actuality, the environment is a dynamic network of living systems that are likewise adapting and evolving. So, Capra (1997: 222) asks, "Which adapts to which?" He answers by quoting Lovelock (1991: 99): "So closely coupled is the evolution of living organisms with the evolution of their environment that together they constitute a single evolutionary process". What appears as adaptation is the on-going co-evolution of interconnected, self-organising systems, as is proposed by Kauffman (1993) with his concept of fitness landscapes, and by Horwitz and Wilcox (2005) in relation to parasitism. However, the terms 'adaptation' and 'evolution' pertain to change. As discussed earlier, from the proposed complexity point of view reality is emerging (i.e. perpetually created by interacting systems) rather than changing. In this context co-evolution is better understood as *co-creation*.

According to Capra (1997), this notion that new life forms are created as living systems *converge* (i.e. couple) is radically different from the conventional neo-Darwinian view, which depicts life unfolding as species *diverge* from one another. He states that it forces biologists to acknowledge the important role of cooperation in evolutionary process (as was recognised by early Russian evolutionary theorists with their notion of *mutual aid*). Margulis and Sagan (1997: 29) similarly argue that cooperation is as important an evolutionary force as competition, stating "Life did not take over the globe by combat but by networking. Life forms multiplied and complexified by co-opting others, not just by killing them". Capra states that Darwinism overlooks the fact that all of nature's struggles occur within a wider context of cooperation, and that organisms network, establish relationships, and live inside one another. This is contrary to the thinking of nineteenth-century social Darwinists who saw nature as being driven by competition, and whose views continue to largely dominate modern western thinking (Capra 1997: 226).

Specifying new territory: a complexity-based epistemology

The Newtonian epistemology flows from the ontology of a material, atomic and objective reality. It is premised on the assumption that there is an external world that is knowable. Knowledge is created as our senses send data about this outside world to our minds. Because this reality is objective and separate from us, it is the same for all observers. Hence, in the Newtonian epistemology, knowledge is an imperfect reflection (or map) of an objective reality, which is a configuration of matter located in absolute space and time. Knowledge gained about any particular system is a reflection of that system's pre-existing order (Heylighen *et al.* 2007). Heylighen *et al.* (*ibid.*) state that because the Newtonian world is simplistic and deterministic, the Newtonian epistemology is also premised on reductionism. Complexity is only apparent. Any given phenomenon, however complex, can be understood by dismantling it and examining the cause-and-effect and relationships between its parts. Since the Newtonian universe is driven by upward causality (where 'things' are causally constructed), time is theoretically reversible. When given sufficient information about a system's present state we can obtain certain knowledge of its past and predict its future (*ibid.*).

The complexity-based ontology outlined in this chapter proposes that there is no external substantial reality. Rather reality is a dynamic immaterial pattern of relationships, emerging in a creative flow of becoming (Whitehead and Sherburne 1981). Because this reality is an interconnected whole, all observers are continuous participators in its creation, making objective observation impossible. There is no external reality waiting to be discovered, as this requires a stable persistent reality existing independently of the observer. This means reality cannot be discovered, only the subjective past can—knowledge is intrinsically subjective, as postulated by postmodern philosophy (Heylighen *et al.* 2007). Capra (2010) notes that while the participator notion is relatively new to modern physics, it has long be a central principle of eastern mysticism, where it is believed that knowledge can only be obtained through full participation, never by observation alone. Indeed, some mystic traditions view observer and subject as inseparable and indistinguishable (*ibid.*), a notion with strong parallels with quantum theory.

In the proposed complexity-based ontology, reality is also understood as a web of nested complex systems. Each complex system, or entity, is in essence an irreducible pattern that has spontaneously emerged from the patterns of relationships created by coupling between its component subsystems. This embeds reality with indeterminate and irreducible processes (as per quantum theory, chaos theory and complexity theory), meaning future events are *constrained* rather than *determined* by past events. Thus, reality is inherently uncertain and unpredictable. *Probabilistic* prediction, however, is often possible due to the formation of stable patterns when entities couple into feedback loops. The irreducibility of complex systems also means they cannot be understood and predicted by simply breaking them down and analysing their component parts, as per Cartesian reductionism. For this reason, complex systems are resistant to reductionist analysis, as this requires some part of the system to be cut off from 'external' feedback loops (i.e. holding other parts of the system constant) so it can be examined in isolation.

Capra (1997) suggests that most reductionists do not understand the limits of reductionism because they do not understand the importance of pattern. Because life is made up of the same atoms and molecules as non-living things, reductionists believe the laws of biology can be reduced to that of chemistry and physics. Capra counters that there is something more to life than just substance, "something non-material and irreducible—a pattern of organisation" (*ibid.*: 81). When a living thing is dissected its components are still there but its pattern of relationships is destroyed; it is now dead (*ibid.*). As per classical thermodynamics, this is also the case for non-living systems. When a non-living system is *closed* (i.e. cut off from interacting with other systems), it eventually succumbs to the forces of entropy and 'winds down' until it becomes completely disordered—it essentially *dies*.

Given these differences, a complexity-based ontology requires an epistemology radically different to that proffered by the Newtonian paradigm. One such epistemology is furnished by Maturana and Varela (1980) in their theory of autopoiesis. From an autopoietic perspective, knowing is an act of relating or coupling with one's surroundings. Through this process reality is *specified*, or created, rather than revealed. In other words, we do not gather information about reality, we *specify* or co-create it via our couplings with other entities. The world we see is "not *the* world but *a* world we bring forth with others" (Maturana and Varela 1992: 245)—reality, with us in it, making itself. This concept will be central to discussions in the next chapter.

Conclusion

In the sections above, I propose a complexity-based paradigm as a more complete alternative to the Newtonian paradigm. Table 5.1 provides a summary comparison of these two paradigms. In the complexity-based ontology there is no external substantial reality. Rather, reality comes forth as an active process of relating. Through this process, patterns of relationship are dynamically emerging from nested sub-patterns of relationships, in a circular feedback process of upward and downward causation, occurring simultaneously across and within system dimensions/levels and between system dimensions/levels. At some level, all things in the universe are eventually connected via this web of relationships. Therefore the universe is a whole, but it is a whole via relationship, not substance. Those things that appear substantial or isolated only do so because of the scale at which we are observing them—substance and separation (i.e. atomism) are illusions created by spatial and temporal scale. The Newtonian view of reality is thus regarded as a 'flattened' (or lower-order) three-dimensional abstraction of a dynamically emerging four-dimensional universe.

It is through *scale* that the paradoxical materialist and idealist notions of the nature of reality become unified in this proposed ontology. At the scale where classic Newtonian mechanics operates, reality can appear and behave as if it is substantial and atomic. At the quantum scale, however, reality is revealed as an incorporeal whole. At certain scales, system behaviours are relatively predictable and immutable, whereas at others they are highly dynamic and subject to unpredictable, radical and spontaneous transformation. Thus, at these scales only probabilistic prediction may apply. At one scale the observer may appear temporally or spatially isolated from other systems, allowing for objective observation. However, when viewed from another scale, the observer is shown to be unified with these systems via higher or lower levels of ordering patterns, indicating the impossibility of holding an absolute objective position. Reductionism (upward causation) can be applied when looking 'downward' at component subsystems, whereas emergence and holism (downward causation) are applied when looking 'upward' at the system whole. In

Table 5.1 A comparison between the mainstream Newtonian paradigm and the complexity-based paradigm proposed in this chapter

Mainstream Newtonian paradigm	*Sources*	*Proposed complexity-based paradigm*	*Sources*
Reality is material and atomic.	*Ancient:* atomists, Aristotle *17th century:* classical mechanics, materialism	**Reality consists of immaterial, dynamic and nested patterns of relationship which form an interconnected whole. Patterns which are persistent in time and space relative to the observer appear substantial.**	*20th century:* quantum theory, process philosophy, holism, general system theory, chaos theory, complexity theory
There is an objective reality that is separate from the observer.	*Ancient:* Aristotle *17th century:* reductionism, materialism, classical mechanics	**Because reality is an interconnected whole, absolute objectivity is not possible. Entities that appear to be isolated at one scale/dimension are connected at others. Reality is subjective and observer dependent.**	*20th century:* quantum theory, process philosophy, chaos theory, general theory of relativity, panpsychism *18th century:* idealism
The arrow of time is reversible. All processes are deterministic and theoretically reversible.	*17th century:* reductionism, determinism, classical mechanics *19th century:* statistical mechanics	**The arrow of time is irreversible. Open systems spontaneously become irreducibly complex over time, and closed systems become irreversibly disordered.**	*20th century:* non-linear thermodynamics, complexity theory, chaos theory, *19th century:* classical thermodynamics
Most natural systems behave linearly and predictably—small changes create small effects, and big changes create big effects.	*17th century:* classical mechanics, linear mathematics	**Most natural systems behave non-linearly because they are coupled through feedback. Small changes can create big effects through positive feedback and sensitivity to initial conditions, and big changes can have little effect due to negative feedback.**	*20th century:* non-linear mathematics, general system theory, cybernetics, chaos theory, non-linear thermodynamics, complexity theory *19th century:* group theory

Table 5.1 Continued

Mainstream Newtonian paradigm	*Sources*	*Proposed complexity-based paradigm*	*Sources*
Atomic reality reconfigures as the result of cause and effect—change is the result of external disturbance.	*Ancient:* atomists *17th century:* reductionism, classical mechanics	**Reality is constantly emerging/being created through self-organisation. Change is internal, the result of perpetual self-creation.**	*20th century:* process philosophy, holism, autopoiesis, complexity theory, chaos theory, non-linear thermodynamics
Reality is reducible—wholes are equal to the sum of their parts.	*17th century:* reductionism, classical mechanics	**Reality is irreducible. Wholes are greater than the sum of their parts. Self-organising systems spontaneously create higher order patterns that are qualitatively different from and therefore cannot be explained by the properties of their parts.**	*20th century:* holism, general systems theory, cybernetics, non-linear thermodynamics, complexity theory, theory of logical types, Gödel's Incompleteness Theorem
Reality is the product of upward causality, therefore there can be no true novelty.	*17th century:* reductionism, determinism, classical mechanics	**Reality is the product of upward + downward (circular) causality via self-organisation. Through emergence, each moment paradoxically has the capacity for infinite novelty while being simultaneously constrained by its spatial and temporal couplings.**	*20th century:* process philosophy, autopoiesis, non-linear thermodynamics, complexity theory, chaos theory
History determines the future. The future is deterministic and predictable.	*17th century:* determinism, reductionism, classical mechanics	**History constrains rather than determines the future. The future is uncertain, and predictability is probabilistic and scale dependent.**	*20th century:* quantum theory, chaos theory, non-linear thermodynamics, process philosophy

Table 5.1 Continued

Mainstream Newtonian paradigm	*Sources*	*Proposed complexity-based paradigm*	*Sources*
Chaos is equivalent to entropy; it is a measure of system's randomness and disorder.	*17th century:* classical thermo-dynamics	**Chaos signals deep, underlying system order that appears as disorder because of the scale of observation. Chaos indicates the system is being observed at a point where it is vacillating between qualitatively different dimensions/ logical levels.**	*20th century:* chaos theory, non-linear thermodynamics
Instability/ chaos in ordered systems is caused by external disturbances.	*17th century:* classical thermo-dynamics	**Chaos is intrinsically embedded in order and vice versa. Chaos and order are co-extant and scale dependent.**	*Ancient:* Pythagoreans *20th century:* chaos theory, complexity theory, non-linear thermodynamics
Space and time are separate and absolute entities—they are the backdrop/ container for the material universe and the immutable laws that govern it.	*Ancient:* Plato *17th century:* classical thermo-dynamics	**Space and time are unified in the fourth dimension as spacetime. Spacetime is not an absolute entity or backdrop to the universe. It is the scope of an emerging, self-organising universe. Laws of the universe are actually large-scale coherent patterns of the universe's self-creation.**	*21st century:* quantum gravity *20th century:* theory of relativity, emergentism, holism, process philosophy, Chew's bootstrap hypothesis *18th century:* Liebniz's relational theory of space and time
Order, complexity, and novelty evolve divergently as the result of natural selection.	*19th century:* Darwinian evolution	**Order, complexity, and novelty emerge spontaneously through self-organisation, i.e. convergence of systems through coupling. Natural selection squashes new order rather than creates it.**	*20th century:* complexity theory (Kauffman), Gaia theory

Table 5.1 Continued

Mainstream Newtonian paradigm	*Sources*	*Proposed complexity-based paradigm*	*Sources*
External reality is knowable. It is revealed to us through objective observation and our senses.	*17th century:* empiricism, positivism *20th century:* logical positivism, scientific realism	**Reality is *specified* through our active participation. The reality we see is not *the* world, but *a* world we bring forth with others.**	*20th century:* quantum theory, autopoiesis, paradigmatic revolutions, postmodern theory *18th century:* idealism

the complexity-based epistemology reality is 'specified' rather than revealed, through one's active participation rather than objective observation.

Richardson (2005) similarly uses complexity-based arguments to propose a philosophy of *quasi-critical pluralism*, which suggests that no one discourse, neither idealist nor materialist, should dominate. In doing so, he essentially tries to dissolve the stalemate between idealists and materialists by deriving a postmodern argument from complexity science. He suggests this philosophy is quite 'empty', not unlike that of Buddhism, in that it does not privilege any one perspective. Heylighen *et al.* (2007) note that while complexity ideas have had impact in fields outside of the 'hard' sciences, such as sociology and organisational management, they have had little penetration in mainstream philosophy, perhaps because of complexity theory's mathematical and computer-based origins. They suggest another reason could be because analytic Western philosophy finds the holism, uncertainty and subjectivity of complexity inimical.

While a paradoxical conception of a unified reality may be largely alien to Western thinking, it has long been a central tenet of Eastern philosophy (Capra 2010). The cardinal Buddhist doctrine of *pratītyasamutpāda* (dependent arising) dictates that all phenomena arise together in a mutually interdependent web of cause and effect. Eastern mystics recognise the individuality of things while being aware that separateness is relative and that apparently opposing and isolated phenomena are actually part of a larger singular one-ness (*ibid.*). Because this notion is difficult to accept in our normal state of consciousness, Zen Buddhists use *koans* to train their minds to be open to this paradox. Koans[22] are mysterious stories that demonstrate the essential dynamic interconnectivity of apparently paradoxical phenomena (Fitzpatrick 2005). Most cannot be resolved by reason or intellect, and thereby force the practitioner into a different level of consciousness or comprehension (Tarrant 2005). Buddhism also posits that Nirvana, the ultimate state of enlightenment, is achieved when the apparent separateness of self and the apparent dualities of the world are revealed as illusions (Capra 2010). Figure 5.4 shows a stroboscopic photo of a woman waving her arms, giving them a near-solid, wing-like appearance. This bears resemblance to familiar Eastern images depicting

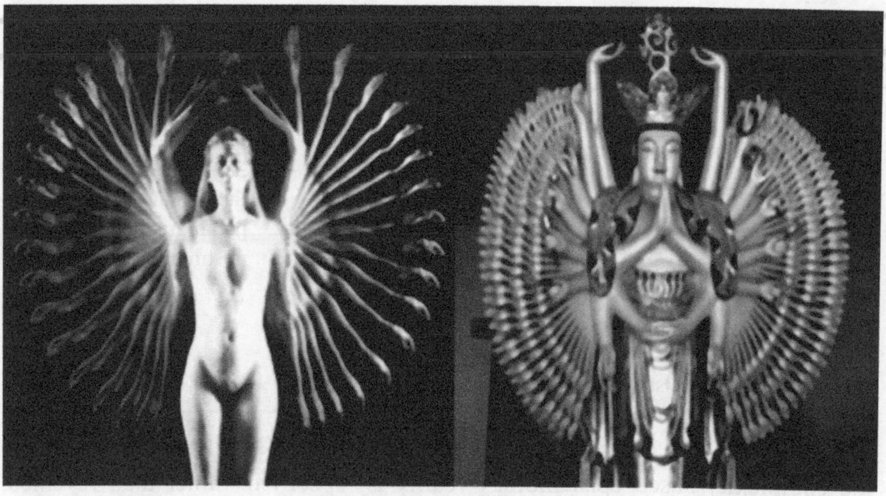

Figure 5.4 Similarities between a stroboscopic photo and a thousand-armed Bodhisattva
Source: stroboscopic photo by Prof. Andrew Davidhazy (retired), School of Arts and Sciences, Rochester Institute of Technology; thousand-armed Bodhisattva by Tengu800, Wikimedia Commons.

Buddha and Hindu gods with multiple arms—further evidence that Eastern mystics grasped the notion of emergent spacetime long before Western physicists did. Perhaps then, given its parallels with Eastern mystical worldviews, a complexity-based ontology presents the possibility of a Western scientific path to Nirvana.

The complexity-based paradigm outlined in this chapter is premised on assumptions about the nature of reality and knowledge which are radically different to those underpinning the mainstream Newtonian paradigm. In Chapter 6, I will show the implications of these complexity-based assumptions on how we have traditionally understood cognition, learning, organising and other social phenomena.

Notes

1 This relates to Cilliers's (1998) notion of *incompressibility* in complex systems.
2 Recognising that many different thinkers and schools of thought played significant roles in the evolution and framing of these ideas.
3 The concept of 'becoming' predates these authors, having also been featured in the works of earlier philosophers such as Hegel (18th–19th centuries) and Heraclitus (500 bc).
4 Other complexity-based paradigms have been put forward by numerous other authors in recent decades, including Morin (2008), Dupuy (1983), Cilliers (1998), Wells (2013) and Mitchell (2009).
5 Bearing in mind that no paradigm is 'true', and that every paradigm is "a tremendously limited understanding of an immense and amazing universe that is far beyond human comprehension" (Meadows 1999: 19). As Cilliers (1998) states, if something is truly complex it cannot be adequately explained by a simple theory.

6 Although Prigogine's views on this are controversial, particularly in relation to quantum physics, which was long believed to be time symmetric, recent research is now indicating a foundation for irreversibility and the arrow of time in quantum settings (e.g. Batalhao *et al.* 2015; Silva *et al.* 2014).

7 Note how photographic images are 'flattened' depictions of reality—the normal photograph of a ball is a two-dimensional depiction of a three-(spatial)-dimensional entity (a ball), and the stroboscopic image is a three-dimensional depiction (two spatial dimensions plus the dimension of time) of a four-dimensional entity (a ball moving through space).

8 This effect is best illustrated by looking at frame-by-frame movement sequences in video format, for example http://moframes.net/collection/baseball/.

9 See Gunderson and Holling's (2002) concept of *panarchy*.

10 Einstein suggested that spacetime is curved by the presence of matter (like a bowling ball placed on a rubber sheet). But this still suggests a pre-existing, albeit dynamic and interacting, spacetime (see Chapter 2). But if spacetime is emergent, perhaps its curvature is a by-product of the 'shape' of reality's emerging patterns. For instance, larger higher-order patterns create large curves, which we perceive as stronger gravity or larger warps in the gravitational field. So in a way, if we flip Einstein's rubber sheet diagrams upside down, we get a better intuition for what's happening—gravity isn't a mysterious field being warped by matter, it is the emerging *shape* of 'matter' (i.e. patterns of interaction).

11 Bear in mind that the transverse sections illustrated in Figure 5.2 are not only slices of space they are also slices of *time*. If one could observe a living human body in real time from the perspective of a transverse section, one would see tissue and cells dynamically interacting.

12 Nor could the contents of the transverse section be predicted or understood from looking at the whole human (without dicing him or her up). They are of different logical orders.

13 As depicted in the video at www.madsci.org/~lynn/VH/transverse.html.

14 The experience would equate to seeing the flowing sequence (e.g. http://moframes.net/collection/baseball) of all events that have occurred in the universe, all at once.

15 Richardson argues that this term is misleading, because 'greater than' connotes a "common measure to compare the whole and its parts and that by this measure the whole is greater than the sum of those parts" (Richardson 2004: 76). He believes this wrong, because wholes are qualitatively different from their parts.

16 This also relates to Gödel's incompleteness theorem and the notion of incompressibility in complex systems (Cilliers 1998; Richardson *et al.* 2001).

17 Watzlawick *et al.* (2011) note that although many theories of change and persistence have been formulated in Western culture, most of these have been about either change or persistence, not the interrelationship between the two.

18 Richardson (2005) notes that a key difference between a complicated system and a complex system is *history*. Whereas a computer or a machine's present state is largely the same as it was a moment ago (on the macro-scale of human observation—on a quantum scale this is no longer true), making its future state easy to predict, a living system's state is impossible to pin down because it is temporally coupled with its history in a way that cannot be observed. This makes its future very difficult to predict.

19 Despite appearances, chaotic behaviour is not the sign of a randomly disordered system (i.e. the entropy that forms in closed systems)—rather it is the product of an underlying pattern of higher dimension order that emerges in open systems far from equilibrium (see Chapter 4).

20 Capra (1997) uses the analogy of a tree to illustrate that how we understand a system depends on how we observe it. When we draw a tree, we might include branches, twigs and leaves, and we depict it is as an isolated, independent entity. However, in our drawings we usually don't include the roots, as we don't see them from our surface perspective. Yet the tree's roots are often as expansive as its above-ground parts, forming an interconnected network with mycorrhizal fungi and the roots of other trees and vegetation, such that there are no boundaries between trees. The patterns of relationships (i.e. objects) that we see are dependent on the observer's perspective and the method of observation.

21 According to Capra (2014), Chew's bootstrap hypothesis is a considerable departure from traditional physics research, which is centred on the search for matter's fundamental constituents. Although popular in the 1970s, by the 1980s Chew's theory had been overshadowed by the standard model of particle physics (which posits the existence of fundamental fields and particles). Capra suggests, however, that Chew's theory may be revived if assumptions about the structure of spacetime are found to be the flaw in string theory.

22 Examples of Buddhist koans include:
- What is the sound of one hand clapping?
- (i) A monk asks Baso, "What is Buddha?" Baso says: "This mind is Buddha". (ii) A monk asks Baso, "What is Buddha?" Baso says: "This mind is not Buddha".

References

Abbott, E. A. (1884). *Flatland: A Romance of Many Dimensions*. London: Seeley & Co. Retrieved from https://books.google.ca/books?id=139bAAAAQAAJ&dq=flatland+a+romance+of+many+dimensions&source=gbs_navlinks_s.

Batalhao, T., Souza, A., Sarthour, R., Oliveira, I., Paternostro, M., Lutz, E. *et al.* (2015). Irreversibility and the arrow of time in a quenched quantum system. *arXiv preprint* arXiv:1502.06704.

Bateson, G. (2000). *Steps to an Ecology of Mind: Collected Essays in Anthropology, Psychiatry, Evolution, and Epistemology*. Chicago, IL: University of Chicago Press.

Bohm, D. (2012). *Wholeness and the Implicate Order*. New York: Routledge.

Bohr, N. (1958). *Physics and Human Knowledge*. New York: John Wiley & Sons.

Borges, J. L. (1998). On exactitude in science. In his *Collected Fictions* (vol. 3; ed., trans. A. Hurley). University of Michigan: Viking.

Briggs, J. and Peat, F. D. (1989). *Turbulent Mirror: An Illustrated Guide to Chaos Theory and the Science of Wholeness*. New York: Harper & Row.

Capra, F. (1997). *The Web of Life*. London: Flamingo.

Capra, F. (2010). *The Tao of Physics*. Boston, MA: Shambhala.

Capra, F. (2014). The unification of physics. Retrieved 15 April 2015 from www.fritjofcapra.net/the-unification-of-physics.

Checkland, P. B. (1981). *Systems Thinking, Systems Practice*. Chichester: Wiley.

Chew, G. F. (1968). 'Bootstrap': A scientific idea? *Science 161*, 762–765.

Cilliers, P. (1998). *Complexity and Postmodernism: Understanding Complex Systems*. London: Routledge.

Crutchfield, J. P., Farmer, D., Packard, N. and Shaw, R. (1986). Chaos. *Scientific American 254*, 46–57.

Dupuy, J.-P. (1983). *Ordres et Désordres: Enquête sur un Nouveau Paradigm*. Paris: Seuil.

Fitzpatrick, S. (2005). Open-ended tangled hierarchies: Zen koans and paradox in public administration. *International Journal of Public Administration 28*(11–12), 957–971.

Foucault, M. (1980). *Power/Knowledge: Selected Interviews and Other Writings, 1972–1977*. New York: Pantheon Books.

Funtowicz, S. O. and Ravetz, J. R. (1993). Science for the post-normal age. *Futures* (September), 739–755.

Gunderson, L. H. and Holling, C. S. (eds). (2002). *Panarchy: Understanding Transformations in Human and Natural Systems*. Washington, DC: Island Press.

Heisenberg, W. (1958). *Physics and Philosophy*. New York: Harper Torchbooks.

Heylighen, F., Cilliers, P. and Gershenson, C. (2007). Complexity and philosophy. In J. Bogg and R. Geyer (eds), *Complexity, Science and Society* (pp. 117–134). New York: Radcliffe Publishing.

Horwitz, P. and Wilcox, B. A. (2005). Parasites, ecosystems and sustainability: an ecological and complex systems perspective. *International Journal for Parasitology 35*(7), 725–732.

Howard, N. (1971). *Paradoxes of Rationality: Theory of Metagames and Political Behavior*. Cambridge, MA: MIT Press.

Jones, A., Pimbert, M., Jiggins, J. and Hall, F. (2010). *Virtuous Circles: Values, Systems and Sustainability*. London: International Institute for Environment and Development.

Kauffman, S. (1993). *The Origins of Order: Self-Organization and Selection in Evolution*. New York: Oxford University Press.

Korzybski, A. (1958). *Science and Sanity: An Introduction to Non-Aristotelian Systems and General Semantics*. New York: Institute of General Semantics.

Laplace, P.-S. (1951). *A Philosophical Essay on Probabilities* (trans. F. W. Truscott and F. L. Emory). New York: Dover.

Lovelock, J. E. (1991). *Healing Gaia: Practical Medicine for the Planet*. New York: Harmony.

Lovelock, J. E. and Margulis, L. (1974). Atmospheric homeostasis by and for the biosphere: the gaia hypothesis. *Tellus 26*(1–2), 2–10.

Machado, A. (1912). Proverbios y cantares. In his *Campos de Castilla*. Madrid: Renacimiento.

Mad Sci Network. (2013). Transverse plane: A guided tour of the visible human. Retrieved 22 October 2013 from www.madsci.org/~lynn/VH/transverse.html.

Mandelbrot, B. B. (1967). How long is the coast of Britain? *Science 156*, 636–638.

Margulis, L. and Sagan, D. (1997). *Microcosmos: Four Billion Years of Evolution from Our Microbial Ancestors*. Berkeley, CA: University of California Press.

Maturana, H. and Varela, F. J. (1980). *Autopoiesis and Cognition: The Realization of the Living*. Dordrecht: Reidel.

Meadows, D. H. (1999). Leverage points to intervene in a system. Retrieved from www.donellameadows.org/archives/leverage-points-places-to-intervene-in-a-system.

Meppem, T. and Bourke, S. (1999). Different ways of knowing: a communicative turn toward sustainability. *Ecological Economics 30*(3), 389–404.

Mitchell, S. (2009). *Unsimple Truths*. Chicago, IL: University of Chicago Press.

Molnar, H. and Meadows, M. (2001). *Songlines to Satellites: Indigenous Communication in Australia, the South Pacific and Canada*. North Melbourne: Pluto Press Australia.

Morin, E. (2008). *On Complexity*. Cresskill, NJ: Hampton Press.

Norris, R. P. and Hamacher, D. W. (2011). Astronomical symbolism in Australian Aboriginal rock art. *Rock Art Research 28*(1).

O'Connor, J. J. and Roberston, E. F. (2003). Jules Henri Poincaré. Retrieved 17 August 2012 from www-history.mcs.st-andrews.ac.uk/Biographies/Poincare.html.

Pagels, H. R. (1985). Is the irreversibility we see a fundamental property of nature? *Physics Today 38*(1), 97–99.
Penrose, R. (1995). Consciousness involves noncomputable ingredients. In J. Brockman (ed.), *The Third Culture: Beyond the Scientific Revolution*. New York: Simon & Schuster.
Prigogine, I. (1997). *The End of Certainty*. New York: Free Press.
Prigogine, I. and Stengers, I. (1984). *Order Out of Chaos: Man's New Dialogue with Nature*. New York: Bantam Books.
Ravetz, J. R. (2005). The post-normal sciences of precaution. *Water Science and Technology 52*(6), 11–17.
Rescher, N. (2008). Process philosophy. *Stanford Encyclopedia of Philosophy* (Summer 2012). Retrieved from http://plato.stanford.edu/archives/sum2012/entries/process-philosophy.
Richardson, K. A. (2004). Systems theory and complexity: Part I. *E:CO 6*(3), 75–79.
Richardson, K. A. (2005). The hegemony of the physical sciences: An exploration in complexity thinking. *Futures 37*(7), 615–653.
Richardson, K. A., Cilliers, P. and Lissack, M. R. (2001). Complexity science: A gray science for the stuff in between. *Emergence 3*(2), 6–18.
Rose, S. (2003). *Lifelines : Life Beyond the Gene*. New York: Oxford University Press, USA.
Rovelli, C. (2003). *Quantum Spacetime: What Do We Know?* Marseille: Centre de Physique Theorique de Luminy.
Silva, J. P., Sarthour, R. S., Souza, A. M., Oliveira, I. S., Goold, J., Modi, K. *et al.* (2014). Experimental demonstration of information to energy conversion in a quantum system at the Landauer Limit. *arXiv preprint* arXiv:1412.6490.
Tarrant, J. (2005). Paradox, breakthrough, and the Zen koan. *Shift: At the Frontiers of Consciousness* (6), 24–27.
Watzlawick, P., Weakland, J. H. and Fisch, R. (2011). *Change: Principles of Problem Formulation and Problem Resolution*. New York: W. W. Norton.
Wells, J. (2013). *Complexity and Sustainability*. New York: Routledge.
Wheeler, A. (1973). From relativity to mutability. In J. Mehra (ed.), *A Physicist's Conception of Nature* (pp. 202–247). Boston, MA: D. Reidel Publishing Company.
Whitehead, A. N. (1948). *Science and the Modern World*. New York: Pelican Mentor Books.
Whitehead, A. N. and Russell, B. (1913). *Principia Mathematica*. Cambridge: Cambridge University Press.
Whitehead, A. N. and Sherburne, D. W. (1981). *A Key to Whitehead's Process and Reality*. Chicago, IL: University of Chicago Press.
Zukav, G. (2001). *Dancing Wu Li Masters: An Overview of the New Physics*. New York: HarperCollins.

6 Through the complexity lens

Cognition, learning, knowledge and organising

What is your aim in philosophy?—To show the fly the way out of the fly-bottle.

(Ludwig Wittgenstein, *Philosophical Investigations*)

Ptolemy and Copernicus skin a cat

The next step in resolving intractable problems involves opening our minds to alternative ways of thinking in light of our revised assumptions, which is the focus of this chapter. The value of opening our minds to new ways of thinking is conveyed by the story of Ptolemy and Copernicus. From ancient Greek times until Copernicus laid out his heliocentric theory, we in the West assumed ourselves to be the centre of the universe, as laid out in Ptolemy's geocentric solar system. The persistence of the geocentric paradigm can be attributed to a single core belief underpinning the religious, cultural and moral fabric of the time: that man is the centre of God's divine creation. However, the paradigm was also buttressed by 1600 years of empirical observations that proved the predictive power of Ptolemy's geocentric theory. This was courtesy of the theory's complex system of epicycles, which mathematically resolved the paradox of retrograde planetary motion created by placing earth in the centre of the solar system (Debus 1978).

Copernicus and Ptolemy's story exemplifies the old adage that there is more than one way to skin a cat. It also brings us back to Korzybski's map versus territory metaphor. Bateson (2000), for example, argues that believing common colds are transmitted when evil spirits are discharged out of a person's mouth and nose during a sneeze then inhaled by others can be just as effective a 'map' (i.e. abstraction of reality) for managing public health as one which views colds being transmitted by microbes. Indeed, the fact that we can devise qualitatively different ways of explaining (i.e. mapping) and predicting the same phenomenon (Maxwell 2000) is why science operates in paradigms, as outlined by Kuhn (1962) and discussed later in Chapter 7. Science is typically a process of searching for evidence to reinforce 'normal' scientific paradigms, rather than a methodical uncovering of reality's 'true' nature. Despite our best intentions to be objective, the paradigm we lend our commitment to is largely an unconscious matter of *faith*.

Today we regard the idea of geocentric universe as absurd. This should not, however, obscure the important lesson served by the history of Ptolemy's and Copernicus's theories. Their story demonstrates how strongly our minds and society shape our perception of reality—as per Kant's transcendental idealism—and how science can be enlisted in shoring up these perceptions. In this chapter I will discuss key phenomena relevant to the problem of knowledge transfer—namely cognition, learning, knowledge and organising—in relation to how they have been understood within the Newtonian paradigm, and then compare this with how they can be alternatively explained from the perspective of a complexity-based paradigm, as synthesised in Chapter 5. Both paradigms clearly have predictive power when describing these phenomena. We will, however, see how the complexity-based paradigm helps resolve some of the learning paradoxes created by the central Newtonian assumption that reality is material, external and independent. This mirrors how the paradox of retrograde planetary motion was resolved when Copernicus challenged the core assumption of Ptolemy's theory: that the earth is the centre of the universe.

Cognition, learning and knowledge

Background

During the 1950s and 1960s, cyberneticists devised information and communication theory to help invent sophisticated *information-processing* machines in the form of digital computers and artificial intelligence. At the same time, Ashby (1952) was developing cybernetic models of the brain, where the brain was conceived as a logical circuit with neurons as base components. In Ashby's view, animals, like machines, behave in predictable ways due to chemical and physical cause and effect (Capra 1997). According to Heylighen *et al.* (2007) research arising from cybernetics has shown that 'intelligence' is generated by patterns of feedback relations that transform sensory inputs into motor outputs. The greater the variety of perturbations a system faces, the greater the variety of compensating feedback loops it will require to maintain cohesion. The system will also require greater 'intelligence' in terms of knowing what compensatory feedback loops to apply in a given situation (*ibid.*). Thus, in the cybernetic view, Cartesian duality is solved, as mind and matter are both simply patterns of relations (*ibid.*). In giving relationship primacy over matter, cybernetics can be seen as more aligned with the ontologies of panpsychism, process philosophy and quantum theory than it is with the material and mechanistic ontologies of Newton and the atomists. Ironically, cybernetic developments in the realm of communication later led to a machine-based conception of the brain and how it works: *cognition*.

For the first half of the twentieth century, studies on learning had focused on observable behaviours and behavioural modification via stimulus and response, conditioning, reinforcement and punishment (Jashapara 2004). This was the realm of behavioural psychology, or behaviourism, which was the principal branch of psychology of the time. In its more radical forms, behaviourism was only concerned with behaviour, and dismissed notions of inward thought and the internal workings of the mind (Hauser 2005). In the 1950s, the cyberneticists' correlations of human intelligence with computer intelligence led to a new *cognitive* understanding of brain function. In this view, the brain is seen as an information-processing machine, much like a computer—rules-based, subject to central logical processing, and possessing local storage (Capra 1997). This led to the *cognitive revolution* initiated by Chomsky (1959), which overturned behaviourism and led to cognitive approaches taking over as the dominant paradigm in psychology. During this time, Broadbent (1958) devised his information processing model of cognition, and Putnam (1960) devised his computational theory of the mind. Neisser (1967) then introduced the term *cognitive psychology*—a field of psychology which views people as dynamic information-processing systems. The computer sciences further reinforced the mind as a computer perspective by using terms such as 'intelligence', 'memory' and 'language' in relation to computers (Capra 1997).

A central notion of cognitive psychology is that individuals construct *mental models* of reality based on their experiences, as first suggested by Craik (1943). These models (or schema) are equivalent to computational structures in a person's working memory that allow them to explore and test different options in their minds prior to taking action (Jones *et al.* 2011, citing others). Learning is understood to occur when a person changes their mental models so they are better able to explain or predict reality. This cybernetically closes the gap between internal and external reality. In the 1970s, psychologist Chris Argyris and philosopher Donald Schön built on Craik's concept of mental models to develop a cognitive theory of learning. They presented two different *theories of action* for explaining people's behaviour:

- the way people *say* they will behave in a given situation (their *espoused theories*); and
- the way they *actually* behave (their *theories-in-use*).

Often, there is a gap between these two theories of action:

> When someone is asked how he would behave under certain circumstances, the answer he usually gives is his espoused theory of action for that situation. This is the theory of action to which he gives allegiance, and which, upon request, he communicates to others. However, the theory that actually governs his actions is this theory-in-use.
>
> (Argyris and Schön 1974: 6–7)

Argyris and Schön (1974, 1996) claim that people's theories-in-use are based on tacit mental maps that govern their actions, often subconsciously. Because these mental maps develop early in life, people are largely unaware they are using them. According to Argyris and Schon, people's effectiveness is improved when their theories-in-use become better aligned with their espoused theories. Improved alignment can be achieved through *reflection*. A person's theory-in-use is supported when the *intended* consequences of an action strategy materialise. When there are *unintended* consequences, there is a mismatch, and this is when learning takes place. Argyris and Schön (1996) suggest there are two possible responses when such a mismatch occurs: *single-loop learning* or *double-loop learning*. Single-loop learning occurs when people try to correct this mismatch by adjusting their actions without questioning their governing variables (i.e. the underlying source of the problem). Double-loop learning occurs when people reflect on and alter their governing variables and change their underlying values. Argyris and Schön (1996) argue that in failing to change an individual's governing values and assumptions, single-loop learning is unlikely to result in real behavioural change. They also observe that most people unconsciously hold theories-in-use that are "systematically counterproductive for double-loop learning, especially when the issues are embarrassing or threatening" (*ibid.*: 76). People are likely to view attempts to bring their attention to such counter-productivity as threatening, and often resort to defensive reasoning and/or avoidance as a consequence. If pressed on the topic, their defensiveness mounts, leading to new 'distortions' in their theory-in-use (*ibid.*). As such, double-loop learning is very difficult to achieve in practice.

The difficulty of achieving double-loop learning is explained in part by *cognitive dissonance theory* (Festinger 1957). When people are presented with new information that conflicts with their existing beliefs or behaviours, they experience an intolerable psychological discomfort termed *cognitive dissonance*. The mind quickly tries to eliminate this discomfort using one of two broad strategies: changing its beliefs and/or actions or resisting the new information. Most knowledge transfer efforts rely on the assumption that people will opt for the former strategy. However, research shows this is often not the case (Sides and Citrin 2007), and, perversely, presenting facts or correcting misinformation can actually *reinforce* people's existing beliefs (Kuklinski and Quirk 2000; Nyhan and Reifler 2010). Indeed, some studies show that the more uninformed (Kuklinski and Quirk 2000), misinformed or ideological (Nyhan and Reifler 2010) people are, the more likely they are to adopt the latter position. This involves trivialising or avoiding the conflicting information, seeking additional support for existing beliefs, and discrediting the information source (Varela 1971). So although cognitive dissonance can lead people to change their beliefs and behaviours, it can also backfire, causing people to reject the new information, shoot the messenger and further entrench their positions.

Despite early findings that brains "operate on the basis of massive complexity, storing information distributively and manifesting a self-organising capacity that is nowhere to be found in computers", the cybernetic notion of cognition quickly swept these arguments aside (Capra 1997). Capra notes that 'cognition as information processing' became the basis of cognitive science and brain research for the following thirty years. He compares the enthusiasm for the 'mind as a computer' metaphor, with the 'body as a clockwork' metaphor that so excited thinkers in the days of Descartes, when clocks—at least as revolutionary as computers in their time—had just been invented (*ibid.*). Stacey *et al.* (2000) sum up the mainstream cognitivist understanding as one that regards brains as passive mirrors of reality—cybernetic, deductive machines that function like computers.

However, there are a number of problems with this mainstream view, which will be discussed in the sections that follow. Given its Newtonian assumptions, the traditional cognitive view of the brain, learning and knowledge is not supported by the complexity-based ontology proposed in Chapter 5. Shotter (1994: 22) states that our notion of mind as something we are in possession of—"an internal, secular organ of thought" that provides an internal representation of an outer external reality—is a myth. In Shotter's view, these ways of thinking about ourselves are so ingrained that it is almost impossible to imagine or speak of ourselves in any other way. Capra (1997) states that while the 'mind as a computer' metaphor was initially a useful framework for scientific understanding, by the 1960s it had hardened into dogma. Neurobiology into the 1970s no longer even questioned the origins and underlying assumptions of 'cognition as information processing'. Capra proposes that the idea of cognition *without* information processing consequently requires a radical expansion of our scientific and philosophical framework. This can be served by a number of alternative theories whereby cognition is viewed as an active process of relating between entities. These theories include: the autopoietic or systems theory of cognition (Maturana and Varela 1980, 1992), the dynamical systems view of embodied cognition (e.g. Varela *et al.* 1991; Wilson and Foglia 2011), and constructivist views of mind and knowledge (e.g. Bahktin 1986; Shotter 1994).

Santiago theory of cognition

In 1987, as part of a contract to the Organization of American States, Humberto Maturana and Francesco Varela published the *Tree of Knowledge* (Maturana and Varela 1992) to address difficulties associated with social communication and knowledge transfer. In the book, Maturana and Varela use their theory of autopoiesis (Maturana and Varela 1980, see Chapter 4) to frame cognition as a continuous process of coupling (feedback) between an organism and other systems that comprise its environment. From this, they make the radical postulation that the process of circular organisation, with or without a nervous system, is identical to the process of cognition. They conclude that all living systems are cognitive systems, and therefore living is the process of cognition. This systems theory of cognition, whereby even the simplest living systems (e.g. bacteria) are deemed capable of cognition, will be henceforth referred to as the *Santiago theory*.

An important aspect of the Santiago theory is that an organism responds to stimuli, or disturbance, *by changing its internal structure* or state, rather than by simply reacting (i.e. cognition is *embodied*). This allows the organism to maintain its circular organisation (i.e. identity) in face of the disturbance.[1] Because the organism is autonomous, it *specifies* its structural response—the disturbance only serves as a trigger. According to Maturana and Varela, organisms only respond to a tiny fraction of the stimuli in their environment. Whether they do so or not is dictated by their structure, which determines what they can perceive. When an organism responds to stimuli by changing its structure, this may in turn cause the source of the stimuli to counter-respond by changing its own structure. This creates a cascade of stimulus-response, which amounts to structural coupling between the two living systems. Behaviour patterns emerge from this coupling, as is similarly proposed in Bateson's (2000) *deutero-learning*.[2] By dynamically recreating themselves with each interaction, together living systems *bring forth a world*. For Maturana and Varela, cognition is an *act of creation*.

Thus, in the Santiago view, cognition is not a process whereby an organism makes a representation of its external environment. Instead, it is a process whereby an organism *co-creates* a world *with* its environment[3] while attempting to maintain its circular organisation. Organisms are constantly immersed in a network of interactions (i.e. cognition). As an organism interacts (i.e. couples) with other systems, together they create a shared "history of evolutionary changes and transformations" through which they become mutually adapted to each other (Proulx 2008: 16). Capra (1997) sums up as follows:

> A living system is a multiply-interconnected network whose components are constantly changing, being transformed and replaced by other components. There is great fluidity and flexibility in this network, which allows the system to respond to disturbances, or stimuli…Certain disturbances trigger specific structural changes, i.e. changes in the connectivity throughout the network. This is a distributive phenomenon. The entire network responds to a selected disturbance by rearranging its patterns of connectivity.
>
> (Capra 1997: 261)

As organisms become more complex, so do their cognitive interactions. This is manifested by the nervous system and brain in higher organisms. In the Santiago theory, having a nervous system[4] serves to enlarge the domain of interactions, the number of relations (i.e. possible behaviours) into which the organism may enter. As such, having a nervous system allows an organism's behaviour to become highly *plastic*. This is consistent with Kauffman's (1993) NK models, which show that increasing connectivity moves complex systems toward the edge of chaos where their responsiveness and capacity for novel behaviour is enhanced. The nervous system therefore does not *create* cognition. Rather cognition is simply the actual *behaving* of the organism in this domain of possible interactions—all knowing is *doing* (Maturana and Varela 1992).

Thus, what traditional cognitive psychologists[5] refer to as a schema, mental model or mind map is not some defining entity in the brain that controls behaviour. Rather, it is a pattern of behaviour that has emerged in relationship with others and the environment, and has become self-reinforcing with repetition over time (Stacey 1996a). The pattern is dynamic, not immutable. However, with repeated iterations it gains cohesion, becoming more stable and difficult to alter. This explains why it is so difficult to provoke double-loop learning in an individual—it is attempting to transform a cohesive pattern of behaviour that has become highly resistant to change. Because this behavioural pattern describes a person's *actual* behaviours, it is essentially equivalent to Argyris and Schön's theory-in-use, but without having to ascribe behaviours to a theory in order to reconcile them with an assumed mental model. Assuming a pattern of relating is a real 'thing'—a mental model of reality stored in the brain—is an error in logical typing, as per the theory of logical types.

Capra (1997) evokes Korzybski's (1958) "the map is not the territory" dictum to sum up Maturana and Varela's view of all entities (living and non-living) as units of interaction: "'no things exist' independent of the process of cognition. There are no objectively existing structures; there is no pre-given territory of which we can make a map—the map-making itself brings forth the features of the territory" (Capra 1997: 264).

Maturana and Varela do not, however, deny the existence of a material world.[6] Rather, they argue that it is in the making instead of being pre-determined: mind and world emerge together. Thus cognition is not a process of *representing* an external world, it is *creating* that world, a process enhanced by language and abstract thought (Maturana and Varela 1992). Their view, that nothing exists independently of cognition, is essentially panpsychic, and comparable to that held by physicists and philosophers such as Spinoza, Whitehead, Jeans and Bohm, all of whom argued that mind and matter are inseparable—that there is no mind without matter and no matter without mind (see Chapter 2). The disparity between the autopoietic and traditional views of cognition is analogous to the difference between the Newtonian and complexity-based understandings of space and time outlined in Chapter 5. In the Newtonian view, space and time are absolute entities that serve as a backdrop for the universe. In the complexity view, unified spacetime is not 'real', as per quantum gravity theories (Wuthrich 2010). Rather, it is the scope (i.e. *pattern*) of dynamic interactions that comprise an ever-emerging reality.

Embodied and enacted cognition

The dominant information-processing view of cognition holds that the "body is peripheral to understanding the nature of mind and cognition" and that "there are no computations without representations" (Wilson and Foglia 2011). Likewise, the connection between the mind and the outside world has traditionally been believed to be of little theoretical importance (Wilson 2002). Cognition is said to occur when the mind manipulates symbols representing the 'real world' to solve problems (Wilson and Foglia 2011). In more recent years, however, there has been a burst of empirical research on *embodied cognition* (e.g. Fischer and Brugger 2011; Koziol et al. 2012). Proponents of embodied cognition (e.g. Clark 1998; Lakoff and Johnson 1999; Varela *et al.* 1991; Wilson and Golonka 2013) disagree with the dominant view, and argue that:

> Many features of cognition are embodied in that they are deeply dependent upon characteristics of the physical body of an agent, such that the agent's beyond-the-brain body plays a significant causal role, or a physically constitutive role, in that agent's cognitive processing.
>
> (Wilson 2002)

In her review, Wilson (*ibid.*) cites a number of authors who have developed dynamical systems theories of embodied cognition, based on the understanding that novel behaviours spontaneously emerge in complex systems. This is also consistent with the Santiago theory of cognition. According to Wilson and Foglia (2011), those with the dynamical systems view minimise (or even deny) the need for a centralised representational processor[7] (as per the traditional cognitive view) to guide an organism's behaviour, arguing that intelligent behaviour emerges through self-organising local interactions in real time. For example, Thelen and Smith (1996) looked at maturation processes in infants and argued that novel behaviours and solutions to problems emerge through the body's activity, without the benefit of pre-programmed cognitive models. A similar perspective is presented by Varela *et al.* (1991), who fused autopoietic notions of cognition with Merleau-Ponty's (1962) phenomenological perspective[8] on cognition, to produce their *enactive* view of cognition. Enactive cognition suggests that one's experience and knowledge of the world arises through coupling of the brain, body and world, through one's bodily engagement with the environment (Varela *et al.* 1991). Thus, enactive cognition is a step beyond the idea of embodied cognition, as it includes an organism's environment as well as its body as part of the cognitive process.

Because they are interaction-centred, the dynamical and Santiago views of cognition are logical extensions of the complexity-based ontology proposed in Chapter 5. This ontology is premised on a reality being perpetually constructed through nested patterns of interactions, one which is ordered in a scale dependent way. Cognition is also nested and scale dependent. We are only consciously aware of things at a particular perceptual level, as dictated by our structure and our circular pattern as a whole (i.e. the whole human level). For example, Beratan (2007) notes that our sensory system is most attuned to those stimuli within the spatial and temporal ranges relevant to individual and group survival. Thus, within the dimension of our potential awareness, we tend to be largely unaware of smaller patterns around us (e.g. ants crawling through the grass at our feet), because they are less likely to interfere with our circular organisation. We are more likely to be aware of larger, higher-order metapatterns that have the potential to disrupt our own circular organisations, such as large animals, other people and the weather. Nonetheless, we are coupling with all of reality, all of the time, at scales beyond our level of conscious perception. Nested within us, our cells and organs are continuously undergoing cognitive processes that we are completely unaware of, interacting with themselves and their external environments to create their own patterns of circular organisation. You cannot 'know' as a cell does, because your emergent cognitive structures are qualitatively different from that of a cell (see the 'darkness principle': Cilliers 1998; Richardson 2004). Likewise, because of the vast interconnectivity of the universe, we are indirectly coupled to the whole of the universe via quantum and cosmic-scale metapatterns we're nested in (as per quantum theory and chaos theory), but completely unaware of.

Through cognition, we bring forth a reality with other entities, which if mapped mathematically represents a subset (or map) of the possible behavioural phase space available to us. This phase space is constrained by the present bodily structure (recognising it reconfigures with every act of cognition) of each interacting entity, and the other systems they are coupling with. The following sections will outline how—from the Santiago perspective—consciousness, self-awareness, behaviour, communication, learning, knowledge and language are all just extensions of cognition—stable higher-order patterns of interaction that emerge from iterative couplings between living systems.

Consciousness: mind and self

Shotter (1994: 45) argues that our current Newtonian world view, where we see ourselves as "all equal, self-enclosed (essentially indistinguishable) atomic individuals, possessing an inner sovereignty, each living their separate lives, all in isolation from each other—the supposed experience of the modern self—is an illusion". An alternative to this way of thinking is the *constructivist perspective*. This view is premised on the work of social psychologists from the 1920s and 30s, namely Mead (1934), Vygotsky (1962) and Bakhtin (1986), who held the view that mind and self are not pre-existing; rather they emerge in social relationship. Mind emerges between an individual and the others he or she is in relationship with—it is *between* them, not individually held *within* them, emerging through the medium of language (Shotter 1994).

The Santiago theory of cognition likewise outlines a process by which the mind and self emerge in relationship. According to Maturana and Varela (1992), when you become aware of something you automatically couple with it, as awareness inevitably triggers internal structural changes within you. *Thus awareness is an act of cognition, an active process of relating.* This is supported by recent neuroscience research indicating that perception is not a passive process, rather it is active and attentional, and occurs through a person's active exploration (Noë and Thompson 2004). In the dynamical systems view, processes rather than structures are the basis of consciousness. Consciousness emerges from large-scale dynamic patterns of activity over multiple frequencies (called 'brain signatures') rather than from specific neuronal circuits[9] (Noë and Thompson 2004, citing others therein).

However, while awareness of one's environment is common to all living thing, *self-awareness*—or consciousness—only appears to occur in higher animals. According to Maturana and Varela (1992), consciousness is meta-awareness—it is awareness that you are aware, knowing that you know. This relates to Jean-Paul Sartre's considered response to Descartes's thesis 'I think therefore I am'. Sartre (1960) argues that the consciousness that says 'I am' is not the consciousness that thinks. When you are aware that you are thinking, that awareness is not part of thinking; it is a qualitatively different dimension of consciousness, of a different logical type (as per the theory of logical types; see Chapter 3). If your mind and self were nothing but thought, you wouldn't know that you were thinking (Tolle 2006). In being conscious, one then makes a distinction between self and others. As the great German philosopher Martin Heidegger (1998: 108) proclaimed, "Transcendence constitutes selfhood".

Consciousness is thus a higher logical order of cognition that emerges from lower order stimulus-response coupling or awareness. It arises from the harmonisation/self-organisation of our living parts (embodied cognition) and the harmonisation between ourselves and our environment (enactive cognition)—our cells, organs, air, water, soil, microbes, other organisms, and so on. Importantly, a being which is self-aware not only couples with its environment to bring forth an external world, but also with itself to bring forth an *internal world of imagination*. This in turn increases the domain and complexity of its potential interactions with the environment (Maturana and Varela 1992), leading to even greater responsiveness and behavioural plasticity, as per Kauffman's (1993) NK models.

Communication and language

In the traditional information processing view of cognition, communication is regarded as information transmission. This differs from Maturana and Varela's (1992) autopoietic perspective. In their view, social phenomena occur when reciprocal coupling between organisms (i.e. cognition) creates a 'third-order' unity, which is a stable pattern of interactions. They refer to the coordinated behaviours that emerge from this third-order unity as *communication*. This means communication is not an exchange of *meanings* (things) derived from information or individual mental models; rather, it emerges from the dynamics of structural coupling. Communication is therefore common to all living things, but becomes more complex as nervous systems become more complex. Maturana and Varela distinguish between *instinctive* and *learned* communication. Instinctive communication is derived from structures that have arisen in an organism's development[10] through the course of its shared evolutionary history. Learned communication arises *between* organisms from their history of social interactions. Because learned communication is specific to the context and social history of the organism, it requires a complex nervous system. Maturana and Varela refer to learned communication as *linguistic*, and note that it is common to many animals.

However, linguistic communication does not become *language* until there is communication about communication—that is, *metacommunication*, at a higher logical level (as per Bateson 2000; see Chapter 3). Capra (1997: 280) illustrates this using an example he was given by Maturana. Every morning a cat meows in the kitchen and its owner gives it milk. This is linguistic communication, a mutual coupling of interactions (the cat signalling, the owner responding). If one morning the owner was to run out of milk and ignore the cat's meowing, and the cat was to say, "Hey, I'm meowing, why haven't you brought my milk?" this would be *language* as it is communication about communication. Once metacommunication is possible, language begins to emerge, allowing for increasingly complex coupling of peoples' thoughts and actions (Maturana and Varela 1992), and therefore even greater behavioural plasticity.

Capra (1997) distils Maturana and Varela's ideas as follows. Language is comprised of words, which are *tokens* for the linguistic coordination of actions, meaning words represent certain structural couplings between entities. Words create the notion of *objects* by linguistically distinguishing a set of relationships from the larger web of dynamic relationships in which they are networked (i.e. their environment). For example:

> at a picnic we can use words as linguistic distinctions to coordinate our action of putting a tablecloth and food on a tree stump. In addition, we can also refer to those linguistic distinctions (in other words, make a distinction of distinctions) by using the word "table" and thus bringing forth an object.
>
> (Capra 1997: 281)

Thus "as language arises, objects also arise as linguistic distinctions of linguistic distinctions" (Maturana and Varela 1992: 210). Note that there is a lack of similarity between the linguistic distinction (e.g. the word 'table') and the behaviour it coordinates (e.g. what we do to distinguish the table—setting the tablecloth and food; Maturana and Varela 1992); the map is not the territory. From our language a hierarchy of logical types emerges.

Elements of this relationship-based understanding of language can be seen in other cultures. For example, anthropologist Jean Briggs, in her interview with CBC radio (Kennedy 2011), notes that the Inuit are more concerned with the relationships between things than they are with 'things' themselves. She recounts how one elder described words in a dictionary as 'dead', because for the Inuit, words without context and personal experience have little use or meaning. That language somehow 'dies' when broken down into words isolated from the relationships of their real-world context reflects the principles of emergence. It also echoes Capra's (1997) observation that dissecting an organism reveals its parts but destroys the patterns of relationship that give it life. Closing off a system from other systems (i.e. controlling 'external' variables) to reveal and control its cause-and-effect machinations leads to the system's 'death' via entropic disordering, as per classical thermodynamics.

A key feature of language is that it allows those who use it to describe themselves and their circumstances using linguistic distinctions. Thus, consciousness (self-awareness) and self-differentiation are essential precursors to language (Maturana and Varela 1992), just as they are for *imagination*, which is *self-projection*. The idea that objects are linguistic distinctions within our consciousness echoes Hegel's absolute idealism, which posits that objects must exist in our minds prior to becoming real and material. Even more so, it echoes Australia's aboriginal creation myths, which, in the words of Chatwin, "tell of the legendary totemic beings who had wandered over the continent in the Dreamtime, singing out the name of everything that crossed their path—birds, animals, plants, rocks, waterholes—and so singing the world into existence" (Chatwin 1987: 2).

Shotter (1994) draws a distinction between *representational-referential* communication and *rhetorical-responsive* communication. Representational-referential communication involves attempting to represent an external reality that will be passively understood by others. This aligns it with the traditional information processing view of cognition. Rhetorical-responsive communication is ordinary everyday conversation, which consists of trying to persuade others to let you do as you wish, or to cooperate with you. In this sense it aligns with the Santiago view of cognition, as rhetoric involves patterns of interaction between parties working to maintain their respective identities. Shotter (1994) notes that although rhetorical-responsive communication is used far more often than representational-referential, it is completely ignored by traditional cognitivists.

Memory and imagination

The Santiago view of cognition is also supported by recent neuroscience findings on memory and imagination. Buckner and Carroll (2007) show that remembering is in essence a constructive, imaginary process, rather than one of retrieving stored memories as it is understood in the information processing view of cognition. Thus, the neural mechanisms involved in remembering an event and imagining the event are the same—we do not reproduce events with our memories, we *construct* them (Schacter and Addis 2000). Every time a memory is activated, it is reconstructed and changed based on the individual's present context, explaining why our recollections of past events are so variable and unreliable (*ibid.*). *Prospection*—the ability to project oneself into an imagined future—involves thinking about the future, remembering the past, and conceiving the viewpoints of others—all of which use the *same* core brain network (Buckner and Carroll 2007). Interestingly, the brain regions activated during prospection mode are identical to those activated when the brain is at 'rest', suggesting that prospection (i.e. imagining) is the brain's default state when not engaged in demanding tasks (*ibid.*). This makes sense from a Santiago perspective, whereby cognition is understood as an organism continuously coupling with itself (body and imagination), other systems around it, and their shared history.

As a consequence, we do not have an unmediated perception of reality—our representations of reality are largely a product of our imagination (Kennedy 2012, citing others therein).[11] Thus, the past is always being reinvented and personal history—who you thought you were and who you think you are now—is changing every moment. We cannot perceive anything in the world without it being shaped by our memories, desires, dreams, etc. Indeed *we imagine the current moment.* Simply imagining an experience changes our brain activation patterns and therefore transforms our reality by changing how we perceive and act into it (Kennedy 2012, citing others).

However, our potential to transform our reality through imagination is constrained because we do not live alone—we live in social and biophysical systems that constrain our imaginations and what is possible (*ibid.*). As such, our imaginations couple us spatially (with other systems) and temporally (with past states) to the present and the past. Our bodies also serve to couple us with the past, through our shared evolutionary history. These couplings shape the present moment and constrain the next by delimiting our phase space, the domain of possible interactions available to us. However, our imaginations also couple us with the future—but in a creative rather than a constraining way. For within the constraints imposed by our past and present lies an infinity of possible future states (as per chaos theory; see Chapter 4). And as we imagine the future, we enact it, our imaginings triggering responses in our bodies, which in turn trigger cascades of stimulus-response between ourselves and our environments, as per the Santiago and dynamical theories of cognition. Thus, we are not *predicting* the future via analysis of information stored in our memories. Rather, we are *creating* the future through our imaginations. Again, this revives Hegel's notion that objects must exist in our consciousness before becoming real and material. If one abandons the determinism of the Newtonian paradigm and accepts instead Prigogine's claim that there is an arrow of time (Chapter 4), that reality is perpetually under construction, then the *future can only ever be imaginary*. In so being, the future *is* imagination—the two are indistinguishable. Hence, the notion of prediction becomes paradoxical. This resonates with Kauffman's (2012) statement that "Consciousness is a participation in The Possible, an ontologically real Res potentia".[12]

Through language, we navigate our constraints and coordinate our imaginations with others. This coordination radically expands the domain and complexity (plasticity) of interactions we can potentially enter into. Hence, it also expands the range of possible futures we can co-create, for better or for worse. Imagination is not neutral—wherever it takes us there will be winners and losers; and our actions do not necessarily bring about the effects we imagined (Kennedy 2012, citing others), given constraints imposed by our histories and the unpredictable effects of our interactions.

Learning and knowledge

Traditional cognitive learning perspectives are premised on the Newtonian assumption of a real, independent world that is external to the learner and about which knowledge can be *acquired* and mapped onto the learner's brain. Knowledge is generated as we observe reality with our senses, then store information about it in our minds as mental representations of the actual world. Decision-making occurs as our brains identify and evaluate if-then hypotheses based on mental models and memories, compare the various alternatives, then make boundedly rational decisions (Beratan 2007). We learn when we use new knowledge to adjust our mental models so they more accurately represent external reality. In so doing, we make better decisions about how to act and reconfigure reality to our advantage, thereby improving our evolutionary fitness (Capra 1997).

Nisbett and Wilson (1977), however, found that people asked to explain their behaviours rarely interrogate their memories to assess how it is they processed information to decide on an action. Rather, they tend to resort to a pool of culturally acceptable explanations, or theories, for how they felt and behaved—the likely equivalent of Argyris and Schön's (1996) *espoused theories*. People apply or generate causal theories about their actions because they do not actually know what makes them behave or think as they do—that is, there is usually no conscious process of evaluating stimuli prior to response (Nisbett and Wilson 1977). Indeed, neurological studies have found that 98 per cent of what the brain does is nonconscious and without conscious guidance, and even our conscious reasoning is largely biased by the unconscious (Beratan 2007, citing others). Recent neuroscience research also calls into question the idea that that our brains store conscious information (Noë and Thompson 2004).

In the traditional cognitive view, learning is believed to occur atomically *within* individuals. An alternative to this view is the social constructivist perspective, whereby learning occurs via the ongoing relationships and interactions *between* individuals (e.g. Gergen 1997; Shotter 1994). Knowledge is continuously *created* between learners through their social interactions and experiences (Ertmer and Newby 1993; Shotter 1994)—there is no storing or retrieving of information in mental models. Maturana and Varela (1992) also reject the idea of knowing as grasping 'facts' about an objective world then storing them in our heads. Knowledge, or learning, is essentially 'successful cognition', whereby an organism couples with other systems without interruption to its own circular pattern of organisation (*ibid.*). We presume to have gained knowledge whenever "we observe an effective (or adequate) behaviour in a given context" (*ibid.*: 174). The regularities we perceive in the world (i.e. knowledge) are simultaneously the result of both our social and biological histories (i.e. they are both biologically and socially constructed). Because we can never experience the world from an independent viewpoint, we can never be certain of our assertions about it (*ibid.*), a notion which is consistent with the observer dependence of reality in the complexity-based ontology outlined in Chapter 5.

Jean Briggs (Kennedy 2011) provides another example from the Inuit that illustrates how some cultures understand learning and knowledge in ways that are more in line with the Santiago and constructivist perspectives. According to the Inuit, if you want to know about 'hunting', you have to go out and hunt; you cannot have meaningful knowledge of it through reading or school. Likewise, you cannot know 'cold' unless you have experienced and had to contend with it in its various forms. As such, Inuit traditionally do not teach their children by spoon-feeding them information; rather, they construct elaborate situations that enable their children to draw their own lessons from various life circumstances. This perspective relates to the *situated learning* and *community of practice*[13] model of how knowledge is constructed, which highlights the limitations of transmitting explicit knowledge abstracted from its real-world context (Brown and Duguid 1991). In this view, training and other methods that isolate knowledge from practice or abstract it from the knower (e.g. office procedures) are regarded as ineffective vehicles for learning (*ibid.*).

Social construction of reality

Because cognition, in the Santiago sense, is understood as a process of interacting with our environments "every act of knowing brings forth a world … all doing is knowing and all knowing is doing" (Maturana and Varela 1992: 26). The idea that we bring forth the world through our interactions with others is also seen in the theories of social constructivists such as Shotter (1994), Gergen (1997) and Bakhtin

(1986). They see mind, self and social phenomena as arising in the ongoing relations between people, rather than arising independently as 'things'. Through their conversations, people both create and are created by their social reality. Shotter regards this as analogous to the processes at work in Prigogine's (1967) dissipative structures: while isolated systems progress towards increasing disorder—as per the second law of thermodynamics—open systems which are in continuous interaction (i.e. conversation) with their surroundings progress towards *increasing* and often transformational order. These open systems not only become transformed, they also transform their environments, creating their own ecological niche to ensure their self maintenance (Shotter 1994). The transformative role of language is also a central theme in postmodern philosophy, as evinced by Foucault (2002: 54) who states: "The task consists of not—or no longer—treating discourses as groups of signs (signifying elements referring to contents or representations) but as practices that systematically form the objects of which they speak".

Bakhtin (1986) argues that all social phenomena are constructed through dialogue, in the ongoing relations between people. He states "Truth is not to be found inside the head of an individual person, it is born between people collectively searching for truth, in the process of their dialogic interaction" (*ibid.*: 110). Through dialogue novelty is produced in an ongoing tension between converging forces of agreement and unity and diverging forces of disagreement and difference. This is consistent with Shotter's (1994) contention that everyday conversation is primarily composed of rhetoric—the arguments people use to persuade others to cooperate with them. Power relations determine whose rhetoric is expressed or suppressed, thereby dictating the emergence of official ideologies (Bahktin 1986). The persuasive power of rhetoric is demonstrated by research showing that *priming* certain patterns of neuronal activation affects the cognitive outcome—for instance, how a question is asked and how alternatives are presented strongly affects decision outcomes (Beratan 2007, citing others). This undermines the notion that people apply logical analysis when making decisions.

Problem formulation and orders of change

Constructivism is melded with mathematical logic in the work of Paul Watzlawick, a psychologist, philosopher, and radical constructivist who followed the footsteps of Gregory Bateson (Bodin 2007). In 1974, Watzlawick joined together with John Weakland and Richard Fisch to publish their classic book *Principles of Problem Formulation and Problem Resolution* (Watzlawick *et al.* 2011). In this work, the authors apply Whitehead and Russell's theory of logical types (1913), and Galois's group theory (see Chapter 3) to psychotherapy. Watzlawick *et al.* suggest that people are often faced with hierarchies of logical types, and the puzzling consequences of confusing levels in these hierarchies. In their analysis of psychotherapeutic interventions, they found that most logical attempts to solve human problems either failed or caused the problem to worsen. The authors looked at logic theories that deal with persistence and change, and how these relate to problem formulation and resolution. From this they discovered that therapeutic approaches focused on creating *first-order change* tended to result in problem perpetuation, whereas those focused on *second-order change* tend to result in problem resolution.

According to Watzlawick *et al.*, first-order change is that which occurs *within* a given system, leaving the system unchanged. Second-order change *changes the system itself.* This distinction can be understood by drawing on cybernetics, group theory and Heraclitus's unity of opposites. Heraclitus's unity of opposites implies that the world of experience is made up of interdependent opposites, from which reality derives substance (which is also reflected in quantum theory). This means that the behaviours of a particular system are the result of a complex circular network of cybernetic feedback loops that regulate the system and maintain homeostasis in the face of change (i.e. causal effects are circular rather than linear). Therefore, as per group theory, changing something into its opposite (opposites being of the same logical type, and therefore of the same system level) creates no overall change in the pattern of the system; rather it just feeds the system with more of the 'same'. Attempting to change a system using first-order change reinforces the system's existing behaviour, which may result in an *impasse* or *deadlock.* It can also escalate the problem by creating a *vicious circle* or a *game without end*[14] (Watzlawick *et al.* 1967), which in turn lead to conflict and conflict engendering 'solutions', such as arms races, etc. This is the basis of why people often feel they are 'banging their head against a wall' when attempting to resolve personal or social issues.

Second-order change is *qualitatively* different from first-order change. Watzlawick *et al.* (2011: 23) describe how second-order change typically appears contrary to common sense, often as "uncontrollable, even incomprehensible, a quantum jump, a sudden illumination". Sometimes it is even interpreted as an act of grace. Although apparently mysterious when viewed from inside the system, it is simply the result of a change in the rules or premises governing the system as a *whole* (meaning change at a higher logical level). The authors metaphorically illustrate the thinking required for second-order change using the classic nine dot problem, where the player is asked to connect all nine dots using just four straight lines, without lifting pen from paper (Figure 6.1).

Most people fail to solve the puzzle, because they make the assumption that the solution must be found within the square (i.e. the 'system'); however the only way to solve the puzzle is to go *outside* of the square (Figure 6.2). Thus, Watzlawick *et al.* (2011: 25) conclude that "failure to resolve a problem doesn't lie in the impossibility of the task, but in the attempted solution". The solution can be found in examining the assumptions that *govern the system as a whole*, not in looking at the *parts* of the system or the rules and assumptions that govern these individual parts. This clearly relates to the theory of logical types, as well as emergence and transformation in holism and complexity theory. Trying to stop a vicious circle or a game without end requires whole system change. Trying to solve these types problems by implementing first-order change either escalates the problem or *is* the problem.

Watzlawick *et al.* observed that second-order change usually occurs spontaneously. People they spoke to who had effected second-order change could not

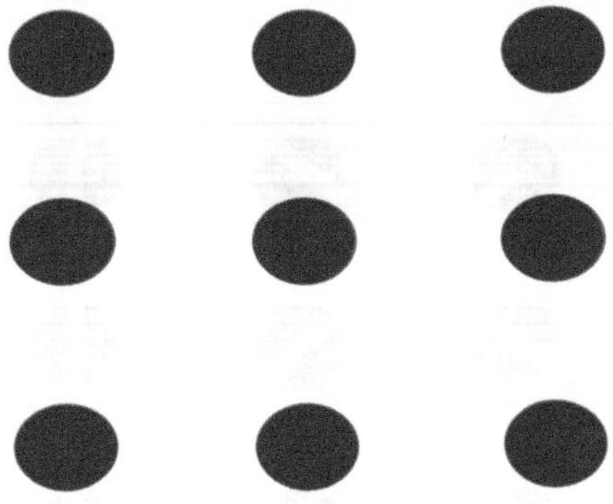

Figure 6.1 Nine dot problem

articulate how it was that they arrived at their interventions. Arduous analysis of the problem at hand creates only first-order change. From their observation, the authors outlined these principles of second-order change:

1 Second-order change applies itself to first-order attempts to solve the problem.
2 Second-order change always appears weird, unexpected or paradoxical.
3 Second-order change deals with the situation in the here and now (what), *not* the presumed causes of the situation (why).
4 Second-order change techniques transcend the self-referential nature of the attempted first-order solution, and put the problem in a different *frame*.

(Watzlawick *et al.* 2011)

Watzlawick *et al.* suggest the fourth principle of re-framing a problem can be used to help achieve second-order change. Reframing means conceptualising the problem at a 'meta' level, or a 'meta-reality', whereby the system whole is viewed from a different perspective. The authors emphasise that this meta-reality is a subjective not an objective reality, quoting Ardrey (1970: 3): "A territory, for example, cannot exist in nature, it exists in the mind of the animal". The authors note that what is 'real' is typically only what a large number of people have agreed to call real (i.e. it is subjective). However, after a time the agreement is forgotten and the agreed reality becomes reified into a fixed *objective* reality in the minds of the people. People categorise the objects and phenomena they encounter as a way of simplifying and making sense of the world's complexity. However, these categories are mental constructs, they are of a different logical type than the actual objects they represent. To confuse our representations of reality with actual reality is an error in logical typing, the equivalent of believing the map *is* the territory (Korzybski 1958). This is why people often fail to see a way out of a problem—they are caged within their own limited constructs of reality, their 'little maps' from which they do not stray (Bateson 2000; Watzlawick *et al.* 2011).

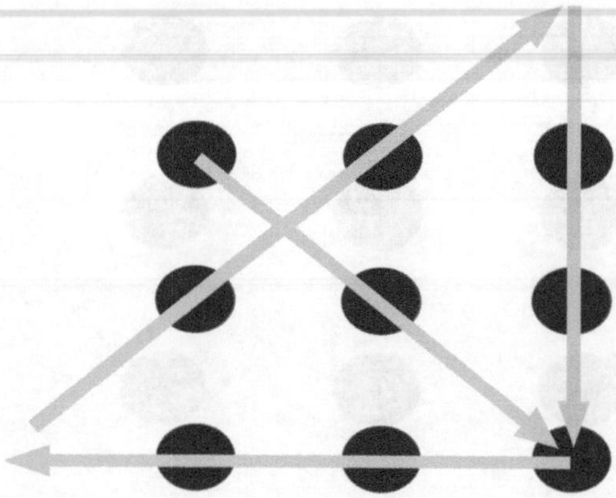

Figure 6.2 Nine dot solution

Analysis versus insight

Watzlawick *et al.*'s theories relate to recent neuroscience research by Kounios *et al.* (2006), which indicates that the brain has two problem solving modes: one that applies conscious analytic reasoning using the left hemisphere, and the other which non-consciously and non-rationally arrives at a sudden *insight* using the right hemisphere.[15] In psychology, insight is said to occur when the solution to a problem emerges suddenly and unexpectedly, in an 'Aha!' moment (Bowden and Jung-Beeman 2003). Watzalwick *et al.*'s (2011) description of the spontaneous occurrence of second-order change and its seemingly weird and paradoxical nature is connected to the phenomena of insight. Kounios *et al.*'s work (2006) shows that creative individuals, in having more diffuse attention that takes in a wider range of sensory inputs, are more likely to solve problems via insight. Conversely, less creative individuals, who have a narrower focus that takes in a smaller range of stimuli, are more likely to solve problems using analytical search strategies.

Insight and analytic strategies are suited to different types of problems, with the brain's right hemisphere playing a special role in solving problems requiring insight (Bowden and Jung-Beeman 2003; Kounios *et al.* 2006). McGilchrist (2012) notes that the left hemisphere of the brain deals in discrete pieces of information, and provides precise, analytic and reductionist thinking. Conversely, the right hemisphere concerns itself with complex pattern recognition, surveying the wider environment (the big picture) for distant connections, integrating and channelling information, and providing holistic, insightful thinking and overall meaning. While the right side of the brain knows what the left is doing, the left hemisphere is rarely aware of what is going on in the right (*ibid.*). Schooler *et al.* (1993) found that insight-based problem solving is impeded by asking subjects to verbally explain their thought-processes when given a problem to solve, as this causes their brain activity to shift to the left hemisphere. Jung-Beeman (quoted in Lehrer 2008) suggests that complex tasks such as language require processing by both hemispheres so that the brain can "see the forest *and* the trees. The right hemisphere is what helps you see the forest".

Kounios *et al.* (2006) found that insight occurs *after* a period of intense focus and attempted problem-solving by the left-hemisphere, during a state of mental relaxation, disorganisation and openness that activates the right hemisphere and allows it to seek out remote connections needed for problem resolution. Sandkühler and Bhattacharya (2008) have likewise found that relaxation makes the brain more open to novel ideas and problem-resolution via insight. Whereas insight is enhanced by a relaxed and wandering mind, focus and pressure to produce can eliminate its possibility—concentration diminishes creativity. Once insight suddenly emerges, the subject reacts both with surprise and certain recognition that they have found the problem's solution, which now appears obvious (Schooler *et al.* 1993; Watzlawick *et al.* 2011). These findings mirror Ghiselin's (1954) compilation of reflections by notable creative geniuses, among them the mathematicians Poincare, Whitehead, and Hadamard. These individuals often reported that their problem solving was not a product of purely conscious calculation. Instead, they often had no idea what prompted the sudden appearance of their brilliant solutions.[16] Alfred Whitehead, for instance, described a "state of imaginative muddled suspense which precedes successful inductive generalization" (Ghiselin 1954: 4–5). Humour also appears to be related to insight. A number of neurological studies have linked activation of the brain's pattern-seeking right hemisphere to humour appreciation (e.g. Fiore and Schooler 1998; Shammi and Stuss 1999). Cave (2005) notes that successful jokes involve incongruities between expectation and result, which sabotages the premises of the expectation[17]. In doing so, humour exposes logical contradictions (i.e. paradox) in our thinking or in certain situations; hence, the important role of satire in exposing social and political problems.

The distinction between analysis and insight can be examined from the complexity viewpoint outlined in Chapter 5. Analysis, or conscious reasoning, works well when dealing with relatively simple problems. However, when dealing with complex, intractable, or wicked problems, making *quantitative* adjustments (e.g. doing more or less of the same) to the interactions between subsystems comprising the problem situation may perversely reinforce the higher order governing pattern (i.e. the attractor) responsible for the undesirable situation. This can result in a game without end or a deadlock, as per Galois's group theory. In addition, ascribing an incorrect logical type to a system's pattern may generate paradoxical solutions that escalate error rather than correct it, such as vicious circles and *double binds*[18] (see Chapter 7), as per the theory of logical types. Examples include the unintended negative consequences of economic or policy decision-making,[19] such as perverse incentives. It also explains Watzlawick *et al.*'s (2011) observations that onerous analysis often fails to effect second-order changes needed to resolve complex problems, and may instead exacerbate the problems at hand. Argyris and Schön's (1996) notion of single-loop learning can be understood as a form of analytical thinking, in that it involves adjusting existing behaviours without challenging a problem's governing variables. This leads to ineffective first-order change, which often causes problems to either fester or escalate (Bartunek and Moch 1987).

In contrast, creative, non-reductionist processes are more likely to trigger insight into higher order patterns that are maintaining the problem situation, and present novel, *qualitatively* different solutions for transforming these patterns (i.e. transforming the system attractor). To change the pattern of the system whole, a *jump* to a qualitatively different, higher order pattern of behaviour is required, one that is not seen or recognisable at the subsystem level (the equivalent of a *step function* in cybernetics). This jump is of critical importance *because it provides a route out of the system* (Watzlawick *et al.* 2011). This reflects Argyris and Schön's (1996) notion of double-loop learning, whereby a problem is addressed by altering its governing variables,[20] leading to second-order change (Bartunek and Moch 1987). Perversely, however, attempts to logically invoke such a transformation in another person often backfire, causing the individual to become defensive and entrench their existing pattern of behaviour (Argyris and Schön 1996), as per the theory of cognitive dissonance (Festinger 1957; Nyhan and Reifler 2010).

From the Santiago perspective we can see that analysis involves consciously imagining the cause-and-effect machinations of a problem. Because imagination is embodied, it triggers action on the basis of what we imagine, and in so doing may actually exacerbate the problem by 'feeding' the feedback loops that cybernetically generate and maintain the problem pattern. Parallels can be found in new understandings of neuroplasticity (e.g. Doidge 2007; Schwartz and Beyette 1996), which propose that dysfunctional behaviours, such as obsessive and compulsive behaviour, arise when people are highly attentive to and over-analyse their problems. In doing so, they repeatedly activate and thereby paradoxically reinforce the coherence of neural patterns generating the problem behaviours (Schwartz and Beyette 1996).

In consciously conjuring and analysing a problem in our imaginations, we also potentially retard our possible range of responses. This is because doing so necessarily pre-supposes a set of possible cause-and-effect relationships, thereby carving out a section of phase space to which the system (i.e. person) then tends to limit itself. By shifting brain activity to the left hemisphere, conscious analytic thinking gets in the way of a person having an unmediated experience of reality via their right hemisphere, which would otherwise open up the responsive possibilities presented by a greater portion of the person/system's phase space. According to Senge *et al.* (2004), seeing a situation as a problem to be analytically resolved, distances us from experiencing how reality is actually arising. Senge *et al.* believe this is related to the notion of *suspension* posited by Varela (Depraz *et al.* 2000), which involves removing oneself from one's habitual stream of thought. Suspension involves viewing our assumptions and thoughts at an arm's length, allowing us to become aware of our thoughts so they have less influence on what we see. This allows us to 'see our seeing', so that we simply notice our thoughts, rather than holding tight to them. By "hanging our assumptions in front of us" (Bohm quoted in Senge *et al.* 2004: 29) suspension enables us to avoid imposing pre-existing mental frameworks and 'groupthink' on what we are seeing (Senge *et al.* 2004).

McGilchrist (2012) argues that Western culture has over-emphasised the conscious reasoning role of the left hemisphere at the expense of the right. As a result, we value precision and quantity over creativity and quality, and have become more inward looking and neglectful of the 'bigger picture' and environmental influences. This contrasts with Eastern philosophies such as Buddhism, which are centred on 'emptying' the mind and breaking free of binding patterns of conscious thought.

Organisational learning and management

Background

Scholars have also applied notions of cognition, learning and knowledge to organisations. Learning and knowledge in an organisational sense have been of keen interest since the early twentieth century, when industrial managers began using Taylor's scientific management[21] to increase industrial production and efficiency (Taylor 1911). This involved stripping knowledge from skilled craftsmen and transferring it to cheap unskilled labour (Drucker 1993; Schutt 2003). The post-industrial economy is referred to by Drucker (1970) as the *knowledge economy,* as it is knowledge rather than craft and industrial technology that drive competitive performance of organisations (Jashapara 2004, citing others). As such, the fields of organisational learning and knowledge management have gained strength over the last 50 years as scholars and practitioners across a variety of disciplines concern themselves with issues around cognition, organisational learning and the effective sharing and application of knowledge.

The organisational learning field is a branch of organisational theory concerned with how organisations learn and adapt. The term was first used in 1963 (Cyert and March 1963), after which its popularity grew dramatically (Crossan and Guatto 1996). Organisational learning theories are largely drawn from theories of individual learning, most notably those rooted in cognitive psychology (e.g. Argyris and Schön 1978). The central tenet of these cognitive theories is a fixed external reality, which is represented and stored in the organisation via cybernetic processes. Hence, the field—as shaped by the 'fathers of organisational learning', Argyris and Schön—is largely premised on the traditional cognitivist belief that people build mental models representing external reality and store them in their minds. Organisational learning scholars frequently refer to organisations as having organisational *schemata* (Bartunek and Moch 1987), or *theories-in-use* (Argyris and Schön 1996). Argyris and Schön (*ibid.*) suggest that organisational theories-in-use are formed from the collective mental models of an organisation's members, which they use to understand the organisation and how they fit into it. The organisation's theory-in-use is responsible for governing its behaviour. Organisational learning happens when a mismatch occurs between the actual and expected results of an action, and a member enquires into it on the organisation's behalf (*ibid.*). Hence, organisational learning occurs by way of information processing, as feedback transforms these organisational theories-in-use by cybernetically closing the gap between external reality and its representation in the minds of organisational members. In the 1990s, Senge (1990) fused aspects of cognitive organisational learning theory with Forrester's systems dynamics (Forrester 1968, 1975, 1993) to derive his concept of the *learning organisation*. In the 1990s, largely in reaction to the dominance of cognitive models in organisational learning, a number of scholars introduced social constructivist approaches to organisational learning (Easterby-Smith *et al.* 2004). In these approaches organisational learning is seen to emerge from social interactions (e.g. situated learning in a *community of practice*; Brown and Duguid 1991; Lave and Wenger 2001).

According to Argyris and Schön (1996), organisations undergo *single-loop* organisational learning when "members of the organization respond to changes in the internal and external environment of the organization by detecting errors which they then correct so as to maintain the central features of theory-in-use". This permits "the organization to carry on its present policies or achieve its present objectives" (*ibid.*: 18), thereby bringing about *first-order* organisational change: that which occurs *within* a given system, leaving the system unchanged (Bartunek and Moch 1987; Watzlawick *et al.* 2011). As such, single-loop learning is a homeostatic process—it works to maintain the present form and functioning of the organisation, although sometimes to the detriment of its long-term survival. In contrast, *double-loop* organisational learning occurs when error is detected and corrected in ways that involve modifying an organisation's underlying norms, policies and objectives (Argyris and Schön 1996)—that is, its schemata (Bartunek and Moch 1987). This brings about *second-order* organisational change: that which changes the system itself, as per the theory of logical types, allowing the organisation to be reflexive and adaptive in uncertain environments (Bartunek and Moch 1987; Watzlawick *et al.* 2011). Single-loop organisational learning is considered useful in relatively stable environments, but potentially disastrous during periods of rapid change. In these unstable environments, single-loop learning actually *inhibits* double-loop learning (Argyris and Schön 1974; Edmondson and Moingeon 1999), resulting in the need for *organisational unlearning* (Nystrom and Starbuck 1984). Training organisational members to be aware of their present schemata so they are better able to change these schemata brings about organisational *meta-learning* (Argyris 2003; Argyris and Schön 1996) and third-order change (Bartunek and Moch 1987).

Knowledge management is a young, multi-disciplinary field, concerned with the generation and sharing of knowledge. It has its roots in the logical behaviourism[22] espoused by Michael Polanyi and Gilbert Ryle in the 1950s and 1960s (Jashapara 2004). Knowledge management scholars draw an important distinction between information and knowledge. This distinction is related to the view that knowledge has two dimensions: *tacit* and *explicit* (Nonaka and Takeuchi 1995; Polanyi 1967), which are the most dominant concepts in the knowledge management literature (Jashapara 2004). Explicit knowledge consists of information that is easily codified and stored with technology (e.g. in databases or reports). However, a considerable component of knowledge goes beyond information to include tacit knowledge. Tacit knowledge is embedded within the minds of individuals and is largely unspoken. It is described by Davenport and Prusak (1997: 5) as "a mix of experiences, values, contextual information, and intuition that provides a framework with which to evaluate and incorporate new experiences and information". Indeed it is through tacit knowledge that an individual or organisation derives its capacity to act (Nonaka and Takeuchi 1995).

Snowden (2003) describes three ages of knowledge management. In the first pre-1995 phase, mainstream theory and practice adopted the Newtonian notion of knowledge as a 'thing', awaiting scientific discovery. During this time, knowledge management focused on using computer technologies to structure the flow of information to decision makers, such as cybernetic process re-engineering and storing information on knowledge databases. The second age of knowledge management was the era when knowledge management as a discipline actually started. Its focus was on improving the productivity of knowledge workers, rather than just storing information on databases. It was sparked by the introduction of Nonaka and Takeuchi's (1995) SECI model, which proposed that knowledge can be converted between tacit and explicit forms through processes of socialising, reconfiguring and learning. Thus, this phase was largely characterised by the search for ways of transforming tacit knowledge into explicit knowledge, and converting it into organisational asset. The third age of knowledge management was characterised by Stacey's (2001) ideas around managing knowledge in complex systems. This approach moves from managing knowledge as a 'thing' that can be disembodied from the worker to managing knowledge as a 'flow', or an active process of relating (Snowden 2003).

The organisational development and management fields are also rooted in Taylor's scientific management. Corporate re-engineering, quality control and assurance, operations research, and total quality management all have their origins in scientific management (Khurana 2009). Cybernetics also had a large influence on these fields. All budgeting, planning and quality control systems are essentially cybernetic processes in that they use negative feedback to regulate activity toward achieving a set goal (Stacey *et al.* 2000). From the 1950s to the 1980s, planned approaches to organisational change—as pioneered by the cyberneticist and social psychologist Kurt Lewin—dominated the management field (Burnes 2005). Lewin (1946) devised *action research* and group dynamics, which together with his theories of planned change are the foundations of organisational development (Burnes 2004). Action research involves iterative cycles of collaborative planning, acting, then reflecting on results. These cycles are essentially cybernetic in that they are intended to be self-regulating. Drucker's (1954) popular management by objective (MBO) approach[23] similarly involves top managers centrally determining organisational goals and strategy, while allowing knowledge workers a level of independence in making and implementing operational decisions aimed at achieving these goals (Waring 1991). Planned approaches to organisational change focus on gradually improving organisational effectiveness using participatory team-based approaches (Burnes 2005). Reductionist methods are applied through the development of organisational objectives and outcomes, with different parts of the organisation incrementally dealing with separate problems and goals one at a time (*ibid.*). In the 1960s, a more elaborate cybernetic approach to analysing and designing organisational processes was devised by Forrester (1968) in the form of *system dynamics*[24] (see Chapter 2).

Until the 1980s organisations were still largely seen as relatively static entities, at or near stable equilibrium—flexibility and rapid responsiveness were not considered critical for their success (Levy 2000). The oil shocks and Western economic recessions of the 1970s, however, exposed the limitations of planned organisational change approaches—they were too cumbersome for a new fast paced world (Burnes 2005). As a result, a number of new approaches to organisational change began to appear, based on ideas borrowed from the fledging complexity sciences. A number of these were centred around *emergent change*—the unplanned, unintentional adaptations and adjustments that organisational members make in their day-to-day work routines (Weick 1979). Among them were Checkland's (1981) soft systems methodology, Weick's (1979) organisational sense-making, and Stacey's (2001) complex responsive processes.

Ralph Stacey is a complexity pioneer in the field of organisational management. Together with Doug Griffin and Patricia Shaw, he established the Complexity and Management Centre at Hertfordshire University in 1995. Stacey and his colleagues posit that the complexity sciences can furnish a more coherent theoretical framework for understanding organisational behaviour. In particular, he advocates a view of organisations as self-organising, complex systems (Stacey 1996a, 2001; Stacey *et al.* 2000), as described in the works of selected complexity scientists, namely Prigogine (1997) and Kauffman (1993). Stacey (2001) also suggests that autopoietic notions of cognition[25] and constructivist views on social phenomena can help build a complexity-based understanding of organisations.

What is an organisation?

According to Stacey (2001), the mainstream organisational literature views organisations as mechanisms or systems—'things' in a Newtonian sense, which learn and generate knowledge in the cognitive sense via information processing. Stacey sums up this view as one where:

> organizations are taken to be learning, knowledge-generating systems of individuals interacting with each other in group/social context…these whole knowledge-generating systems must be managed in some way in order to optimize, or at least improve, their functioning.
>
> (Stacey 2001: 24)

Given its Newtonian assumptions, this cognitive view of organisations is not supported by the complexity-based ontology proposed in Chapter 5. As is the case at the individual level, there are a number of problems and paradoxes associated with an information processing view of organisational learning, knowledge and change, which will be discussed in the sections that follow.

When viewed in the context of Chapter 5's complexity-based ontology, an organisation is understood as a cohesive higher order pattern of interaction emerging from the interactions of its members and its couplings with other systems. Through its interactions, an organisation both shapes and is shaped by its members (which are themselves lower order sub-patterns of interaction). Stacey describes this self-organising process as "large numbers of agents interacting locally with each other … [to] produce overall coherence without any overall plan" (Stacey 2005: 21). Through its interactions, the organisation also simultaneously shapes and is shaped by the other systems (which are also just patterns of interaction) that comprise its environment. While organisations are traditionally understood as *having* 'routines', which are described as dynamic patterns of behaviour (Jashapara 2004), in the complexity view, the organisation *is* a dynamic pattern of behaviour. Likewise, the term *organisational culture*, is just another way of describing persistent patterns of interaction that *are* the organisation.

This understanding is reflected in Stacey's (2001, 2010) view of organisations as *complex responsive processes*—patterns of everyday interaction between people, or the temporal processes of human relating. Stacey's perspective does not abstract people from their interactions in that it does not regard the product of their interaction as being a 'system', i.e. some external 'thing'. Rather, the product of their interaction is simply more interaction between people. These iterative processes of relating are simultaneously cooperative and conflicting, and subject to power differentials between organisational members. It is through these everyday processes of relating that people in organisations cope with uncertainty and perpetually co-create the future together (Stacey 2001). Thus, from a complexity perspective, an organisation is a coherent pattern of interactions *co-creating* reality with other systems (which themselves are just dynamic patterns of interaction), rather than a 'thing' or a system of things moving about in a pre-given external reality. Organising can therefore be viewed as an *extension* of cognition in the Santiago sense, whereby living systems are seen to continuously *bring forth a world* through their interactions (Maturana and Varela 1980, 1992).

Organisational learning

Because the cognitive approach to organisational learning attempts to reconcile with the notion of mental models, its criteria for what constitutes organisational learning includes two elements:

- behavioural change that represents a change in the organisation's theory-in-use (i.e. the collective mental model of its members); and
- demonstration that the knowledge acquired through learning has been 'embodied' in the organisation in the form of artefacts such as memories, policies, and procedures.

(Argyris and Schön 1996)

These artefacts are regarded as the equivalent of an organisational nervous system. They 'store' the organisation's knowledge (Popper and Lipshitz 1998), and serve as the learning mechanisms by which the organisation collects, stores, distributes and uses information (Friedman *et al.* 2005). This creates the paradox whereby the organisation is considered both a *representation* of knowledge (in its behaviours) and a *container* of knowledge (in its artefacts); thus it is both a 'process' and a 'thing' (Lipshitz 2000). According to Lipshitz and his colleagues (Friedman *et al.* 2005; Lipshitz 2000; Popper and Lipshitz 1998), the extension of individual cognition models to organisations also raises the following questions:

- How is learning at the individual level transformed into learning at the organisational level?
- In what ways, if at all, are organisations capable of learning?
- How can managers induce organisational learning in real life?

While Argyris and Schön's theory is consistent with the complexity perspective in its focus on learning as an active process, it is inconsistent in its reliance on mental models. If one dispenses with the assumption that cognition is a matter of information processing using mental models, as does the Santiago theory of cognition, then attempting to insert 'mental models' into the process of organisational learning, as Argyris and Schön do, is unnecessary. The organisation *is* its pattern of behaviour, which is functionally equivalent to its theory-in-use, but without circuitous reference to a theory or model. The notion of mental models becomes both an unnecessary complication and an error in logical typing, as outlined in the discussion around cognition earlier in the chapter. Cognition, communication, consciousness, imagining, dialogue, learning, organising, and all other social phenomena, are simply ascending and nested dimensions of order emerging from increasingly complex patterns of interaction—i.e. they are the same process, just at different levels of complexity. This reflects Stacey's (2010) assertion that individuals, groups, organisations and society are all phenomena of the same ontological level.

In the Santiago view, all living systems are perpetually learning, transforming themselves with every interaction (as per Maturana and Varela 1980). An organisation is thereby *continuously learning* by virtue of its definition, in its continuous interactions with other systems. Hence, learning at the organisational level does not require physical manifestations in the form of artefacts—it is evident in the ongoing patterns of interaction that constitute the organisation's day-to-day behaviour. As is the case at the individual level, organisational learning is seen as *successful* interaction, where the organisation interacts with other systems without disruption to its circular pattern of organisation/identity. This is an adjustment to the traditional Newtonian idea of an organisation *adapting* to its environment, which assumes a background world to which the organisation, as an autonomous entity, must adapt. From a complexity-based perspective, the organisation can be understood as *co-creating the world* with other entities, rather than *adapting to a world* and competing with other entities for a space in it. This complexity view nonetheless maintains a strong 'survival' element: each organisational entity works to maintain its circular organisation/identity while interacting with others, a process subject to power relations, conflict and cooperation.

When viewed from a complexity perspective, it is inappropriate to think of policies, procedures and other artefacts as being the equivalent of an organisational nervous system where knowledge is stored. This is because the role of a nervous system is *not* storing information, as understood by the mainstream information-processing view of cognition. Rather, the role of the nervous system is to increase the number and complexity of interactions an organism can enter into, therefore expanding the organism's domain of possible behaviours, as per the Santiago view of cognition. As such, gossip at morning tea, email and the Internet (as a means of interacting) would be better analogues for an organisational nervous system.

Thus, from the proposed complexity-based understanding, knowledge *is* learning; it is an active process of relating, not a *product* of learning.[26] In sum, it is a *process of co-creating reality* with other entities, rather than a *representation of external reality*. If knowledge is no longer regarded as a 'thing', it can be neither a representation of an organisation (rather, it *is* the organisation—a dynamic pattern of interaction), nor contained within organisation. Again, this is consistent with Stacey's complex responsive processes approach, which sees minds and knowledge as being continually produced in an active process of relating; there is no storing or retrieving of information in mental models (Stacey 2001, 2010). Social constructivist approaches to organisational learning—whereby learning is seen to be socially constructed in the relationships between people (e.g. Lave and Wenger 2001; Wenger 1998)—are also more in line with a complexity based understanding of organisational learning than is the traditional cognitive perspective.

The proposed complexity-based understanding of organisational learning also resolves the question of how individual learning is transformed into learning at the organisational level. The transformation takes place via emergence, when the interactions of individuals self-organise into a higher order unifying pattern constituting the organisation. Because an organisation is of a higher logical order than its members, its behaviours and learnings are not simply a sum of that of its members. Rather, they are emergent, and therefore qualitatively different to that of its members. In other words, an organisation will learn and behave in ways that cannot be predicted from the learning and behaviour of its members. To assume so would be an error in logical typing.

The shadow organisation

A complexity-based understanding of organising can also help explain why attempts at bringing about organisational change so often fail, as discussed in Chapter 1. Once a circular pattern of organisation has become established, it is self-reinforcing and serves only to maintain itself, typically becoming more entrenched over time. This feature of complex self-organising systems helps explain why organisational members become attached to patterns of behaviour that (from management's perspective) have outlived their effectiveness, as exemplified by 'competence traps' (Levitt and March 1988) and conservative processes perpetuated by 'webs of interest' (Argyris and Schön 1996). It also explains why people resist double-loop learning, and become defensive when changes to entrenched organisational patterns are attempted (i.e. second-order change; *ibid.*). In addition, it bears upon Stacey's (1996a) notion of the *shadow organisation.*

Stacey (*ibid.*) observes that all organisations carry out a set of primary tasks via the cooperative efforts of their members—this is the purpose of an organisation's formal, or legitimate, system, which consists of its official policies, procedures, organisational hierarchy, budgets and job titles, all of which tend to be rational, explicit and written down. However, running parallel to this formal system is the organisation's *shadow side*: the "disagreeable, messy, crazy and opaque aspects of your organisation's personality" (Tate 2005: 22). The shadow organisation consists of all the real-time social and political interactions that occur outside the organisation's formal system (Stacey 1996a). Elements of this shadow system include: trust, friendships, ambition, jealousy, fear, insecurity, power struggles, and gossip (Tate 2005). Whereas an organisation's formal system is organised by those in authority, the shadow system is dynamic, self-organising, tacit and obscure; it is political and is often undiscussed or un-discussable (Tate 2005, citing others). The shadow organisation has a life of its own. It is self-serving, governed by the collective personal needs of its members (Stacey 1996a). It can be chaotic, but also very resistant to change, sabotaging any attempts to alter the status quo (*ibid.*). Thus the shadow organisation represents the *actual* behaviour of an organisation, as opposed to how it is *intended* by management to behave.

From a complexity perspective, this means that an organisation may realistically "traverse a substantial phase space around an attractor, which may represent multiple organizational states" far from its formal mission and goals (Levy 2000: 82). The shadow system is often the source of an organisation's failure to efficiently deliver on its mandate (Tate 2005). This is particularly the case when an organisation's shadow system is widely out of alignment with its formal system. On the flip side, the shadow system is the organisation's fount of creativity and innovation (*ibid.*). Because the shadow system is where things *really* happen in an organisation (as opposed to where/how they're *designed* to happen), when the formal system becomes overly bureaucratic or dysfunctional the shadow system can deliver outcomes in spite of it (*ibid.*). Stacey (1996a) suggests that the shadow system is the generator of the mess and disorder needed for learning and ensuring a system doesn't become 'ossified'. As such, Tate (2005) argues that any change initiative that fails to gain the support of the shadow organisation will likely fail.

The hopeful map

This idea of an organisation having two such sides is also reflected in Argyris and Schön's (1996) notion of an organisation having both an *espoused theory* of action and an *actual* theory-in-use—the theory-in-use being equivalent to the organisation's shadow system, or actual behaviour, and the espoused theory being equivalent to the organisation's legitimate system. Indeed, the formal system, or espoused theory, can be understood metaphorically as a *hopeful map:* a 'map' in that it is an abstraction meant to describe the 'territory' (the *actual* pattern of interactions that constitute the organisations), and 'hopeful' in that it describes how managers have *presumed* or *intended* that territory. The 'shadow organisation' and 'theory-in-use' concepts capture the understanding that our 'maps' are often widely out of alignment with the territory they're meant to represent. In the case of organisations, the misalignment is due to our underlying Newtonian assumptions that complex self-organising systems are 'things' predictably subject to cause-and-effect, and which can therefore be designed and controlled to meet externally designated purposes (Stacey *et al.* 2000). And when a formal system does not work as designed—and indeed begins to create new problems or exacerbate the ones it was designed to solve[27]—the organisation's members are stuck with it, making ordinary every day decisions to work around it and still get things done (*ibid.*).

Stacey's notion of the shadow system resonates in the maxim of the famed cyberneticist Stafford Beer (2002): "the purpose of the system is what it does". This reflects his observation that a system's *effective* purpose is often far removed from its *official* purpose. Thus, if the purpose of a system is what it *does*, then the 'purpose' of an organisation, or any other system, is inevitably *emergent* (the result of the interactions between its components and with other systems), rather than *prescribed* by some entity or authority standing outside of it. In this sense, 'purpose' really means 'pattern', or habit, and is an inappropriate term in relation to self-organising systems, as it is teleological in its implication of design, intention and/or fate.

Organisational change

The problem solving discussion earlier in this chapter can also be applied when thinking about change in organisations. Because organisations are circular self-reinforcing patterns of interaction, trying to solve organisational problems analytically—by making quantitative adjustments to the interactions between the organisation's members (doing more or less of the same)—tends to only effect *first-order change* (as per Watzlawick *et al.* 2011). First-order change reinforces the existing pattern of the organisation, and therefore perpetuates the problem at hand, as per Galois's group theory, and Argyris and Schön's (1996) single-loop learning. However, any prescribed attempts at changing the higher order governing pattern of the organisation, i.e. *second-order change*, will likely be greeted with resistance and distrust, as members defensively mobilise to snuff out threatening changes, behaving as what Stacey (1996a) calls 'organisational antibodies'. This effect has been likewise described in relation to double-loop learning (Argyris and Schön 1996), cognitive backfire (Nyhan and Reifler 2010), learning organisations (Senge 1990), and systems dynamics (Meadows 1999). Instigating second-order change may become more difficult as an organisation becomes more complex, and is therefore increasingly governed by higher-order patterns that are far beyond any individual's control.

As a means of countering this effect, several writers (Lipshitz 2000; Argyris and Schön 1996; Argyris 2003; Visser 2007) put forward the notion of organisational *meta-learning*. Meta-learning is said to occur when members of an organisation learn to reflect on their organisation's schemata (or theory-in-use) as a means of detecting and correcting errors and effecting second-order change (i.e. members learn to undertake double-loop learning on behalf of the organisation). However, from a complexity perspective this is a case of confusing logical types (as per Chapter 3). Organisational meta-learning and double-loop learning cannot be undertaken at the level of individuals. For organisations to be capable of meta-learning consciousness (i.e. self-awareness), and some form of meta-communication (language) would be required at the *organisational* scale; allowing the organisation to deliberatively reflect on itself and its governing premises. While a human organisation can be understood as being aware, as are all organisms, *reflection* requires self-awareness (meta-awareness, or consciousness), which only occurs in higher animals and is unlikely to be present in human organisations.[28] Thus, the complexity perspective also resolves the question of what kinds of learning organisations are capable of.

Nor is it likely that organisational resistance to change can be overcome by increasing managerial control. As Stacey *et al.* (2000) note, because the organisation's governing pattern emerges from a multitude of interactions between its members and other systems, the actions of one or a few people in management—even if they hold disproportionate power—are unlikely to effect the desired change. According to Richardson (2005), change in complex systems is very difficult to predict and sometimes undesirable:

> The overall system's behaviour might be radically affected, or the system might absorb any attempt to change and continue relatively unaffected. Though, it is important to remember that attempts to change the system (failed or not) may result in delayed changes despite no apparent immediate reaction. Who knows what chain of events might have been triggered.
>
> (Richardson 2005: 624)

Indeed, managerial attempts to 'fix' certain organisational behaviours may actually inadvertently create serious problems, such as vicious circles and deadlocks, where issues were previously mild or non-existent (Watzlawick *et al.* 2011). This effect has been similarly observed by systems theorists such as Meadows (1999). She states that managers have difficulty discerning which way to push system 'levers' to effect desired organisational change. Leverage points are frequently counterintuitive, and worsen the situation if pushed in the wrong direction. These difficulties are summed up by Forrester (1993), who states:

> First, most difficulties are internally caused, even though there is an overwhelming and misleading tendency to blame troubles on outside forces. Second, the actions that people know they are taking, usually in the belief that the actions are a solution to difficulties, are often the cause of the problems being experienced. Third, the very nature of the dynamic feedback structure of a social system tends to mislead people into taking ineffective and even counterproductive action. Fourth, people are sufficiently clear and correct about the reasons for local decision-making—they know what information is

> available and how that information is used in deciding on action. But, people often do not understand correctly what overall behaviour will result from the complex interconnections of known local actions.
>
> (Forrester 1993: 9)

This relates to the question of how managers can induce organisational learning. Assertions that people can be trained or coached to induce double-loop organisational learning (as per Argyris and Schön 1996; Senge 1990) and second- or third-order organisational change (as per Bartunek and Moch 1987) seem problematic, given the deep complexity of dynamic interactions that are involved in creating a pattern of organisational behaviour to begin with, and the uniqueness of every organisation. Despite their adherence to such assertions, Bartunek and Moch (1987) observe that second-order change usually only occurs 'naturally' (i.e. spontaneously), and often in response to a crisis which 'unfreezes' the existing organisational schemata. Numerous other authors have likewise indicated that crisis, or some form of organisational 'unlearning', is usually needed to precipitate double-loop learning and second-order change (Argyris and Schön 1996; Nystrom and Starbuck 1984; Watzlawick *et al.* 2011). Indeed, Bartunek and Moch even go so far as to suggest that change agents try to precipitate a crisis in order to prime an organisation for second-order change. This makes some sense when viewed from a complexity perspective. Self-organising agents in a complex system simultaneously and paradoxically bring forth both persistence—due to constraints imposed by coupling with past states and shared histories with other systems—and potential transformation—due to amplifying feedback and spontaneous emergence of novel behaviours. As a system in crisis becomes increasingly unstable (i.e. approaches the edge of chaos), it becomes sensitive to amplification of small effects, poising the system for radical transformational change, or collapse into disorder (as outlined in Chapter 4). This paradoxical situation is reflected in the Mandarin symbol for crisis, which is comprised of two characters, one representing danger, the other representing opportunity.[29]

Stacey (1996a) discusses this feature of human organisations in relation to Kauffman's (1993) NK models of complex self-organising systems. When connectivity, diversity and rates of information flow between organisational members and/or outside groups are low, organisations are relatively stable and unchanging. These organisations require little energy or information flow to sustain them. However, they are not really learning or adapting because existing patterns of connection have locked-in, or *ossified*.[30] In order for an ossified organisation to become receptive to change, connectivity, diversity and rates of information flow between people need to increase. This begins to destabilise the organisation, causing it to spontaneously become more responsive and adaptive to its environment. In doing so, novel patterns of organisational behaviour (i.e. new basins of attraction) may begin to emerge. This effect appears to be subject to diminishing returns (Yuan and McKelvey 2004), however, and in the case of too much connectivity, the organisation may alternatively be driven to collapse (Stacey 1996a).

For example, Tan's (2013) review of organisational change shows that organisations with a more diverse workforce have better problem-solving skills and can understand and penetrate more markets. However, the review also found that too much workforce diversity can also create confusion and chaos. Maintaining existing patterns of behaviour that exploit old certainties are seen as beneficial when an organisation is operating in a relatively stable environment (Holland 1975; March 1991). However, in rapidly changing and unpredictable (i.e. turbulent) environments, hanging onto old patterns can be detrimental—the organisation must become more responsive to exploring and adapting to new possibilities if it is to survive (Holland 1975; March 1991; Stacey 1995). This perhaps explains why solutions are short lived in dynamic environments (as per Lipshitz 2000) and why stable solutions breed new problems, (as per Bartunek and Moch 1987), as is widely noted in the ecological resilience literature (e.g. Allison and Hobbs 2006; Berkes and Turner 2006).

Managerial control

Stacey *et al.* (2000) posit that today's organisational management is premised on scientific management (i.e. Taylorism) and systems thinking. They believe this has led to a Newtonian view of organisations as 'mechanisms' (in the case of Taylor, Drucker, etc.), or self-regulating systems that have been externally engineered (in the case of Forrester, Argyris and Schön, Nonaka and Takeuchi, Senge, etc.), which can be studied to reveal the cause-and-effect laws that govern their behaviour. It has also led to the view that managers and researchers stand outside their organisations as objective observers (Stacey *et al.* 2000), and that change in a system is brought about causally, by one entity impacting an another (Shotter and Tsoukas 2011: 334, citing others). Thus, armed with the correct information, managers have the power (by designing rules, procedures, strategies, etc.) to manipulate and control organisations to achieve chosen goals and induce optimal patterns of behaviour. As such, the manager is the 'scientist' who designs the organisational system and determines the rules of interaction for its component parts, and the organisation is the 'phenomenon' under study (Stacey *et al.* 2000).

Stacey *et al.* (*ibid.*) argue that these deep-seated Newtonian beliefs are why managers so often feel they are in charge but not in control, why the systems they design so seldom work, and why planned attempts at organisational change are so often met with frustration. Designed systems, rules and procedures cannot cope with all the eventualities of a self-creating, perpetually emerging entity that is in constant flux (i.e. an organisation), and the turbulent world it is enmeshed in and helping create. As soon as some system mechanism has been identified and described, the organisation has moved on. The Newtonian view of organisations as mechanisms also neglects the role of workers' free choice in shaping organisations (Stacey *et al.* 2000), and fails to recognise organisational change as spontaneously emerging from workers' ongoing interactions (Shotter and Tsoukas 2011). Stacey *et al.*'s (2000) comparisons of the mainstream Newtonian view of organisations with a complexity-based view are outlined in Table 6.1. They argue that

Table 6.1 A comparison of how organisations are understood from a complexity perspective versus the dominant management discourse

Organisations as complex systems	*Dominant management discourse*
1 **Long-term predictability of complex organisational processes is severely limited, because emergent behaviours in complex systems cannot be predicted in advance. Creativity and uncertainty are linked.**	The manager's role is to reduce uncertainty.
2 **Transformational change in organisations occurs via the dynamic self-organising interactions of their members. Creativity occurs through cooperative relationships and the constraints imposed on these relationships by power, politics and conflict.**	Competition is more important than cooperation.
3 **There are limits to individual choice. The organisation's creative future is the unpredictable result of the dynamics of interaction rather than the choice of individuals and small groups.**	Managers have the power to choose and control their organisation's future.
4 **Organisational stability results from the conflicting constraints (power) imposed by relationships, redundancy, and operating at the edge of chaos, where extinction events tend to be small.**	Efficiency and managerial control are sources of organisational stability.
5 **Diversity and difference are sources of organizational novelty.**	Harmony and consensus need to be emphasised.
6 **Ability to plan and design change is limited. Organisations have internal capacity to change spontaneously. Change cannot be designed, planned or optimized.**	Management's role is to design and plan for optimal outcomes.
7 **Organisations have the potential (not guarantee) to succeed and change in novel ways when they are simultaneously stable and unstable.**	Success is equated with stability.
8 **Organisations and their members are driven by their need to express their identity and difference. Goals relating to profit and competiveness are subservient to this need.**	Performance is the all-important motivating force behind an organisation.

Source: adapted from Stacey *et al.* (2000).

understanding organisations as complex systems means accepting the following: uncertainty is inescapable and success can never be guaranteed; cooperation is more important than competition; no individuals, including managers, have the power to control or design organisational change; efficiency destroys organisational stability; diversity and conflict foster innovation; potential success lies in co-existent stability and instability; and organisations are driven by their need to express identity rather than profit or competitiveness.

According to Stacey *et al.* (2000), what *can* cope with a turbulent world are the people who make up an organisation, as they have the capacity to exercise free will and adapt to changing circumstances. They are the reason why organisations get things done anyways, despite the failures of managerial plans and strategies to achieve intended outcomes. Organisational change is not caused by chance or managerial decisions. Rather, it emerges from the interactions and relationships between people in the organisations. Stacey *et al.* further suggest that:

> The notion that managers can choose what happens is so deeply ingrained that it leads to the typical response that if they can't choose outcomes, at least they can choose the numbers and strengths of connections, the qualities of relationship that produce the dynamics of edge of chaos where creative change is possible. However, this misses the point, because no agent is choosing the numbers or strengths of connections for other agents in the system, or for themselves either; even if they were, this is not enough to determine the dynamic.
>
> (Stacey *et al.* 2000: 112–113)

This relates to the increasing application of complexity ideas to finding new ways of strategically manipulating organisations and their agents (see Ahrweiler 2013; Anderson 1999; Baumann and Siggelkow 2011; Buchanan 2011; March 1991). Examples include: enabling evolution of self-organised solutions to problems, altering fitness landscapes for local agents, and reconfiguring or 'tuning' organisational architecture to promote adaptation and innovation. Stacey and others (e.g. Shackley *et al.* 1996; Stacey *et al.* 2000; Stacey 2010), however, criticise complexity-based approaches attempting to improve prediction and control of system behaviour as simply being methodological extensions of more traditional systems-based approaches. Specifically, Stacey *et al.* (2000) dispute a number of assumptions held by other complexity management writers, namely that managers can:

- select simple rules that will create the desired pattern of outcomes;
- create harmony and sharing in an organisation (which ignores the role of conflicting constraints in organisational creativity and stability); and
- learn how complex systems are ordered then use this information to manipulate the system.

They argue that attempting to apply these assumptions will result in frustration as managers attempt to assert control over that which is uncontrollable, a view that is shared by Shackley *et al.* (1996), who paint a picture of institutions *pretending* to manage complexity. In a similar vein, Levy (2000) cautions that the idea that organisations can be 'tuned' to be more innovative is problematic, and questions whether it is possible to know when an organisation is on the edge of chaos, or what control parameters can be adjusted to get it there.

Conversation as currency of change

Given the challenge to prediction and organisational control presented by the complexity perspective, Burnes contests its usefulness, arguing that if organisations are to be understood as:

> dynamic non-linear systems capable of continuous transformation through self-organization, advocates of this approach will need to show either that it is more than just a metaphorical device, or that even as such it is able to resolve the problems of managing and changing organizations more effectively than other approaches that are on offer.
>
> (Burnes 2005: 87)

Stacey (2003), however, cautions against viewing self-organisation as a new way of ordering an organisation or a new form of organisational behaviour to be strived for, as this seems to "assume that self-organisation is some new form of behaviour rather than a different way of understanding how people have always behaved". According to Stacey (2010: 44), when "one focuses attention on the 'system,' one tends to lose sight of the centrality of the process of interaction, which perpetually constructs itself as continuity and transformation". He argues that instead of trying to manage and control the whole system, the emphasis should be on paying attention to one's own local participation, and on patterns of interaction (Stacey 2001). From a methodological perspective, this means that organisational managers and researchers should adopt the role of participative inquirers. This relates to the work of Senge (1990: 69), who sees people in organisations as having the capacity to shift from being "helpless reactors" to "active participants" in shaping their reality and the future. Senge places a strong emphasis on the role of conversation, or dialogue, as it is conceived by the physicist David Bohm (Bohm *et al.* 1991), where a group becomes open to the flow of a 'larger intelligence', and thought is approached largely as collective phenomenon (Smith 2001). This also relates to social constructivist perspectives premised on the notion that if people start talking differently about the world, they will change it; hence, generating change is a matter of shifting conversations (Shotter and Tsoukas 2011, citing others).

Conversation can, however, foster persistence as well as change. Dialogue is the creative exploration of complex issues through active listening and the suspension of personal views and judgement (Bohm 1994; Bohm *et al.* 1991). As such, it can potentially foster conditions for insight and second-order transformational change when dealing with complex problems (Senge *et al.* 2004). Discussion, on the other hand, is the search for best arguments through the presentation and defence of different views, with the participant focused on 'winning' (Jashapara 2004). Consequently, attempting to problem-solve via discussion often reinforces existing power differentials and patterns of behaviour (i.e. first-order change), as members work to defend and entrench their positions (as per Argyris and Schön 1996). Discussion is also often the mode of analytic conversations, whereby issues are rigorously dissected. As detailed earlier in this chapter, focusing on the parts of the problem may reinforce higher order patterns of behaviour that created the problem to begin with (as per Watzlawick *et al.* 2011). In essence, discussion is about trying to dominate the discourse in order to control how the external world is presented and understood. Dialogue, however, is about looking forward, imagining and co-creating a future together. Jashapara's (2004) comparison of these two modes of discourse is summarised in Table 6.2.

Stacey (1996b), however, suggests that dialogue will only change an organisation's dominant pattern of behaviour if its effects on people's actions are somehow amplified. If amplification does begin to occur, a period of discussion and power manoeuvring inside the organisation's shadow system will follow. The more stable the organisation, the stronger its internal 'organisational antibodies'—those members who attempt to block change and protect the status quo (as per Brown and Duguid 2000; Stacey 1996a), and who are often the most powerful constituents of the organisation's legitimate system. The less stable (i.e. closer to the edge of chaos) the organisation is, as during a crisis, the less likely it is that the 'antibodies' will succeed in damping amplification of novel change (*ibid.*). The persistence of novel patterns of organisational behaviour also depends on the cooperative and competitive interactions between other organisations.

Shaw (2002) states that managers need to accept that our interactions are always evolving in unpredictable, uncontrollable ways (despite attempts at sophisticated planning) and recognise the constant and recurring potential for change as we interact. This means appreciating that it is not possible for one individual or group to dictate what behaviour patterns will emerge as a result of dialogue, or any other attempted intervention. However, as described earlier, an organisation's behaviour may become more plastic and responsive to change as the number and complexity of interactions it and its members enter into increase. As such, Stacey and his colleagues argue that rich, diverse and unconstrained conversation between people is the currency of change in human systems (Shaw 2002; Stacey 2001). By increasing conversation between its members and outside groups, the system's ability to respond and adapt to change is likewise increased,[31] and the domain of possible interactions the organisation can potentially enter into are expanded. Shaw (2002)

Table 6.2 Characteristics of dialogue versus discussion

Dialogue	*Discussion*
Purpose: to go beyond one's understanding and become an observer of one's own thinking	**Purpose:** to find the best view and arguments to support the decision that needs to be made
Requires: letting go of power differentials, treating each other equally; exploring assumptions behind closely held views	**Requires:** presentation and defence of arguments (note, however, that power, rhetoric and emotion, rather than logic or good arguments, can win)
Useful for: divergent thinking, trying achieve a richer understanding of the issues rather than fostering agreement	**Useful for:** convergent thinking, fostering agreement, decision-making

Source: adapted from Jashapara (2004).

states that our organising patterns change as we change our patterns of accounting to one another through conversation. Because conversation is integrated into people's day-to-day interactions, it directly shapes their behaviour in the here and now, and thus the reality they are co-creating. Hence, conversation has a more immediate influence on a system's patterns of interaction than do abstract discussions carried out in isolation of ordinary practice and requiring formal pathways of planning and implementation (*ibid.*). Shaw's perspective is echoed in Shotter and Tsoukas's (2011) *ecological approach* to language-based organisational change. According to Shotter and Tsoukas, organisational change occurs spontaneously, as people begin to notice new possibilities through their ordinary conversations and then respond bodily to these events by reorienting themselves towards others and their environments. The authors argue that these sorts of changes cannot be generated by "intellectually devised plans, procedures or protocols; they cannot be done, intentionally, by people taking deliberate actions", (*ibid.*: 346). They draw on the logic of Wittgenstein (1980) in their distinction between difficulties of the *intellect* versus difficulties of the *will*:

> We can formulate difficulties of the intellect as *problems* which, with the aid of clever theories, we can solve by the use of reasoning. Difficulties of the will, however, are quite different. For they are to do with how we *orient* ourselves bodily towards events occurring around us, how we relate ourselves to them, the ways in which we see them, hear them experience them, value them—for it is these *ways* that determine, that 'give shape to', the lines of action we further *resolve* on carrying out.
>
> (Shotter and Tsoukas 2011: 345)

Thus, rather than thinking about change as a mechanical, one-way, cause and effect process, Shotter and Tsoukas see change as emerging from the "two-way *spontaneous responsivity* of living forms" (*ibid.*: 346). All our actions are effectively *interactions*. As such, their perspective, along with Shaw's, aligns with Maturana and Varela's autopoietic view of people bringing forth (or specifying) a world through their ongoing conversations and social interactions.

Written versus verbal dialogue

So what then is the role of the *written* word? Stacey (2005) claims that writing and printing technologies led to emergence of new social phenomena, such as bureaucracies, regulations, management by objective, and wide spread reliance on so-called tools of communication, such as reports, plans, databases, etc. According to Donaldson (2005), writing is not just a means of recording information and articulating and offering our thoughts to the reader; it also forces us to reflect on and organise our own thinking. Thus, writing allows us to enter into a form of dialogue, both with ourselves and with the words of others we would otherwise never meet—like those of Plato written 2000 years ago (*ibid.*). As such, writing proffers the possibility of a much wider audience for our thoughts and imaginings, in both space and time, than that offered by conversation.

Verbal dialogue, especially face-to-face, is characterised by a creative and uninterrupted flow of cyclical stimulus-response between the parties involved, one that has an immediate effect on their patterns of interaction. Written dialogue (excluding near-instantaneous forms of written dialogue, such as text messaging and email), on the other hand, involves long delays between the time of writing and time of reading, and is often a unidirectional exchange. The writer may influence the thinking and behaviours of the reader through his or her words, but the reader's thoughts and reactions are not typically reflected back to the writer, at least not directly. As such, written communication and other unidirectional media, such as TV and radio, lack the creative flow of face-to-face dialogue, and the immediacy of physical response.

Donaldson (*ibid.*) suggests that modern reliance on the written word has cultivated generalised forms of thinking, whereby concepts (e.g. 'strategy' or 'culture') are decontextualised from human interaction. She cites Ong, who states: "Writing fosters abstractions that disengage knowledge from the arena where human beings struggle with one another" (Ong 2002: 43). In addition, Donaldson states that reading and writing are typically solitary activities that distance us from others. She argues that by focusing on written communication, we become distanced from our immediate situations and lose sight of the importance of our working relationships. As a result, we privilege planning over improvisation and engagement, abstract categorisation over direct experience, written records over conversation, and communication tools over human relationships (Donaldson 2005). From a complexity perspective this means we become alienated from each other, distancing ourselves from the immediacy of human interaction from which reality emerges, and the possibility of joint action and novel outcomes.

Vision versus representation

Imagination and vision also play an important role in the relative impact of our communications on the emerging pattern of reality. For example, in his interview with CBC radio (Kennedy 2012), politician Joe Mihevc states that politics is a struggle over people's imaginations. He notes that politics has shifted away from trying to win over the public with detailed policy documents to focusing on *vision* instead. He argues that appealing to people's reason will not get you as far as appealing to their vision:

> Vision is the most important thing in politics, as shown throughout human history. Leaders that have had an imaginative vision have been the most influential. Those that capture people's aspirations unlock our imagination of the world that could be, apart from the world that is.
>
> (Mihevc in Kennedy 2012)

Thus, persuasion is more easily achieved by means of imagination or narrative, than by logic or analysis. This is consistent with the discussion earlier in this chapter, which showed how imagination is a form of self-projection, which continuously transforms our reality by directing how we perceive and act into it. In sharing our imaginings during our conversations we expand the domain of our possible interactions and the possible futures we collectively make. Together, we constantly create and improvise our existence (Kennedy 2012, citing others), in a dynamic tension of agreement and disagreement, cooperation and conflict (as per Bahktin 1986; Stacey *et al.* 2000). As such, our social reality can be understood as an emerging pattern of jostling, rhetoric-mediated imaginings.

This also explains why *rhetorical-responsive* speech is much more common in day-to-day interaction than is *representational-referential* communication (as per Shotter 1994). Rhetorical speech is the medium of jointly envisioning and creating reality in the here and now, whereas representational speech is a passive reflection of a system state put forward for logical analysis. When we consciously apply ourselves to analysing a task, we essentially turn off our creative imaginations and largely disengage from the active process of co-creating reality.

Knowledge management

As touched on previously, mainstream thinking also applies a cognitive view of knowledge, whereby knowledge is a product of learning, stored in people's minds as mental models representing an objective reality. Capra (1997) notes that in this computer-based model of cognition, knowledge is seen as context and value-free, and based on abstract data. Knowledge is regarded as a 'thing' which one can be in possession of, be it tacit or explicit, know 'how' or know 'what', or regarded as a system which can be externally manipulated or controlled. Although those in the knowledge management field have long recognised that knowledge which leads to action tends to be embodied unconsciously within the knowledge holder—as tacit knowledge or know-how (Nonaka and Takeuchi 1995; Polanyi 1967), or within the relationships between people (i.e. relational knowledge; Park 1999)—the field has nonetheless largely focused on finding ways of stripping this knowledge from knowledge workers (Waring 1991) and converting it into corporate asset (Snowden 2003).

From the complexity-based understanding outlined earlier in this chapter, however, knowledge *is* learning—an active *process* of relating, not a *product* of learning that represents external reality. According to Stacey (2001, 2010), knowledge is always a relational process; it cannot then be externally designed or manipulated, stored or managed. He also states that explicit and tacit knowledge are facets of the same communicative process, and hence cannot be discussed separately, or converted from one to the other. Furthermore, intellectual capital cannot be measured, the cleverness of the intellectual elite is irrelevant (it is their interactions that matter), and quality assurance in staff training and development is largely a charade that provokes more anxiety and frustration (Stacey 2001). Rather, learning occurs and knowledge arises in the ordinary everyday interaction between everyone in the organisation, and with those in other organisations (Stacey 2001). Stacey's view resonates with Maturana and Varela's (1992) assertion that there is no knowledge without action and no action without knowledge. It also relates to Shotter's (1994) constructivist view of knowledge as something continuously created between learners through their social interactions and experiences. Shotter and Tsoukas (2011) suggest that this notion of 'relationality', and the emergent uniqueness it bestows on organisational behaviour, is one which mainstream organisational management has found difficult, not least because it means there can be no truly 'transferable' knowledge. As Stacey argues (2001), the knowledge assets of an organisation are the patterns of relationship between its members, assets that are destroyed when these relationships are disrupted.[32]

Hence, from the proposed complexity perspective, a system's relative level of 'knowledge' is a function of the domain and complexity of its interactions, which are what confer its behaviours with plasticity. The knowledge of an organisation is not a measure of the information held within the heads of its members, or within artefacts such as databases, policies and procedures. Rather, it relates to the number and diversity of its connections between the organisation's members and other systems. The 'knowledge' of a system can then be said to increase when new connections/relationships emerge between its agents and other systems. These added connections expand the scope of possible futures the organisation can *co-create,* internally and with other systems. This contrasts with the Newtonian information processing view of knowledge as an asset that helps the organisation *adapt to* an external, pre-given environment.

According to Stacey (2001) many organisations could do more with less, and save time, money and frustration by abandoning ineffective knowledge management efforts that attempt to control or create knowledge as a 'thing'. Instead, any attempts to improve knowledge-creating capacity require attention to the quality and dynamics of human interaction *in the present.* This involves focusing attention on: patterns of power relations and ideologies; who is talking and who is being silenced; who is excluded and who is included; how people manage their anxiety; and whether and how conversations are spontaneous and excited, or dull and repetitive (*ibid.*). Shaw (2002) notes that this requires recognising that conversing *is* organising, and that ordinary free-flowing conversation has the potential to generate transformational change.

Societal change

The complexity-based perspective proposed in Chapter 5 can also be extended to the societal level. Laszlo and Laszlo (2002) apply *General Evolutionary Theory,* Ervin Laszlo's (1996) synthesis of various complexity-based theories, to explain societal phenomena. They suggest that the tendency for open systems to become increasingly complex, as per non-linear thermodynamics, is demonstrated in how human social systems are progressively converging to higher and higher levels of organisational complexity as they become more interconnected. The last few thousand years saw rapid convergence of tribes and villages into ethnic communities, which in turn converged into states, then into nations, and now into large political and trade blocks. Each new level of organisational order emerges out of a period of instability, at which point a societal 'bifurcation', or transition, takes place. These transitions may be smooth, chaotic or catastrophic. During these times, "the reins of power change hands, systems of law and order are overthrown, and new movements and ideas surface and gain momentum" (Laszlo and Laszlo 2002). These events may stabilise into a 'new order', in which the convergent subsystems (e.g. villages, or states) are unified into a new higher order pattern of behaviour; however, they may also cause a collapse into disaggregated individually stable subsystems (*ibid.*).

Davidow (2011) suggests that the Internet has resulted in an over-connected global society, which is becoming increasingly destabilised from the amplifying feedback loops that are emerging as a result. He attributes over-connection to global economic instability and stock market volatility. Similarly, Stark (2006) suggests that the global political system is being driven into a chaotic phase by increasing interconnectivity within and between societies. He suggests a bifurcation point is looming, whereby a new unifying system of global governance will emerge, or catastrophic socio-ecological collapse will occur. This perhaps explains the feeling many of us share, that our worlds are becomingly overwhelmingly complicated. As society becomes more connected, higher order organisational patterns begin to emerge 'above us', and in turn exert increasing 'downward' governing pressure onto our own localised behaviours. Hence, for some of us, our feelings of being out of control mount as these larger, higher order patterns entrain our individual local patterns for their own self-reinforcement. For others, the sense is one of excitement, of being part of something bigger and newer, the twenty-first-century zeitgeist.

Conclusion

In sections above, I outline how mainstream notions of cognition, learning and organising are based on core assumptions of the Newtonian paradigm, then present some alternative understandings of these phenomena that are better aligned with the complexity-based paradigm synthesised in Chapter 5. A comparison of these perspectives is summarised in Table 6.3. Clearly, replacing assumptions of the mainstream Newtonian paradigm with those of a complexity-based paradigm has profound implications for how we understand individual and organisational learning and behaviour. Traditional information processing views of cognition, by using terms such as 'mental model' to describe coherent patterns of behaviour[33] (thereby assuming that the brain actually has stored within it a 'map' representing external reality) are committing an error in logical typing, as per the theory of logical types. The linguistic symbol used to represent, or distinguish, a pattern of relating (i.e. mental model) is assumed to be a real 'thing'. Hence, the map is being confused with territory—a process or pattern is being confused with a 'thing', wherein the two are of clearly distinct logical types.

By understanding cognition in terms of dynamic patterns of interaction, complexity-based understandings of cognition, such as the Santiago theory, avoid this error in logical typing. They are also able to resolve paradoxes created by the

Table 6.3 A comparison of learning, knowledge and organising, as understood in the mainstream Newtonian paradigm versus the proposed complexity-based paradigm

Mainstream Newtonian paradigm	*Sources*	*Proposed complexity-based paradigm*	*Sources*
Our brains store mental models that represent external reality. Remembering is an act of retrieving these stored models.	Materialism, cybernetics, cognitive psychology	**Our brains do not possess storage systems or mental models. Remembering is part of *prospection*, an imaginary, constructive process that couples us with other systems and our shared history.**	Neuroscience (Buckner and Carroll), transcendental idealism (Kant), autopoiesis (Maturana and Varela)
Cognition is information processing: comparing our experiences with our stored mental models, and testing different options in our minds prior to taking action.	Materialism, cybernetics, cognitive psychology	**Cognition is a continuous process of coupling between ourselves and other systems—all living systems are cognitive systems.**	Panpsychism (Whitehead), autopoiesis (Maturana and Varela), dynamical systems view of enacted cognition (Varela *et al.)*
Knowledge is created as we observe reality through our senses, then store it in our minds as a mental representation. We learn when we use new knowledge to adjust our mental models to be a more accurate representation of external reality.	Materialism, empiricism, positivism, cybernetics, cognitive psychology	**Knowledge, or learning, is 'successful cognition', whereby an organism couples with other systems without interruption to its own circular pattern of organisations. Knowledge is not a 'thing'.**	Autopoiesis (Maturana and Varela), dynamical systems view of enacted cognition (Varela *et al.)*
Our mental models of reality dictate our behaviour via information processing.	Cybernetics, cognitive psychology	**Our behaviours are patterns of interaction/ coupling that emerge through our shared history with other systems.**	Autopoiesis (Maturana and Varela), dynamical systems view of enacted cognition (Varela *et al.)*, social constructivism (Shotter, Bakhtin)

Table 6.3 Continued

Mainstream Newtonian paradigm	*Sources*	*Proposed complexity-based paradigm*	*Sources*
Cognition is the means by which we navigate and adapt to an external reality.	Materialism, cybernetics, cognitive psychology	**Cognition is the means by which we mutually construct reality via our interactions with other systems.**	Autopoiesis (Maturana and Varela), social constructivism (Shotter, Bakhtin), process philosophy (Whitehead)
Communication and language involve transmitting information/ messages for processing by the recipient.	Cybernetics, communication theory, cognitive psychology	**Consciousness, behaviour, communication, and language are all just extensions of cognition—higher-order patterns of interaction of that emerge from iterative couplings between living systems.**	Autopoiesis (Maturana and Varela), social constructivism (Shotter, Bakhtin)
By allowing us to exchange information, communication and language help us improve our mental models of external reality and thereby adapt to it.	Materialism, cybernetics, cognitive psychology	**With each new cognitive order (i.e. consciousness, language), the complexity of interactions into which an organism enters increases dramatically, increasing its responsiveness and creative capacity/ plasticity.**	Autopoiesis (Maturana and Varela), complexity theory (Kauffman)
Given sufficient and accurate information, analysis can solve all problems.	Materialism, determinism, reductionism, scientific realism, positivism	**Insight is needed to solve complex problems because analysis cannot quantify across system boundaries. Analysis also impairs insight, and tends to worsen complex problems.**	Neuroscience (Kounios *et al.*), single vs. double-loop learning (Argyris and Schon), first vs. second order change (Watzlawick *et al.*), Theory of Logical types (Whitehead and Russell, Bateson), Group Theory (Galois), chaos theory

Table 6.3 Continued

Mainstream Newtonian paradigm	*Sources*	*Proposed complexity-based paradigm*	*Sources*
An organisation is a 'thing'.	Materialism, organisational learning, organisational management	**An organisation is a higher order unifying (circular) pattern emerging from the interactions of its members and shaped by its couplings with other system.**	Autopoiesis (Maturana and Varela), complex responsive processes (Stacey)
Organisational behaviour is dictated by the organisation's theory-in-use, which is a shared mental model of reality held by its members.	Organisational learning (Argyris and Schon)	**An organisation's behaviour is a governing pattern emerging from a multitude of interactions between its members and other systems; i.e. an organisation *is* its behaviour.**	Autopoiesis (Maturana and Varela), social constructivism (Shotter, Bakhtin), complex responsive processes (Stacey), complexity theory (Kauffman)
Organisations can be manipulated and controlled by those who stand outside of them, in order to achieve externally dictated goals (e.g. profit, competitiveness).	Organisational management, systems dynamics, organisational development	**An organisation's only 'motive' is to maintain its identity—hence it resists changes that might affect its overall governing pattern. It cannot be manipulated or controlled by any one individual or group. Its behaviours are often far out of alignment with its formal purpose. Transformational change occurs spontaneously, often precipitated by crisis.**	Autopoiesis (Maturana and Varela), complex responsive processes (Stacey), complexity theory (Kauffman), organisational learning (Nystrom and Starbuck), orders of change (Bartunek and Moch)
Organisations learn when their members detect discrepancies between the organisation's theory-in-use and reality, then successfully correct the discrepancy on the organisation's behalf.	Organisational learning (Argyris and Schon)	**Organisations are continuously learning by virtue of their ongoing interactions with other systems.**	Autopoiesis (Maturana and Varela), complex responsive processes (Stacey), complexity theory (Kauffman)

Table 6.3 Continued

Mainstream Newtonian paradigm	*Sources*	*Proposed complexity-based paradigm*	*Sources*
Organisational knowledge is a 'thing' that can be stripped from workers and commoditised into an organizational asset. It is embedded into the organisation's policies and structures, and the mental models of its members.	Scientific management (Taylor), knowledge management (Drucker; Nonaka and Takeuchi), organisational learning (Argyris and Schon)	**Organisational knowledge is not a 'thing' that can be stripped from its members; rather it relates to the connectedness of the organisation and the domain and complexity of interactions it can potentially enter into.**	Autopoiesis (Maturana and Varela), complex responsive processes (Stacey), complexity theory (Kauffman)
Organisations improve their adaptive capacity by improving their knowledge about their environment.	Organisational learning (Argyris and Schon; March), knowledge management (Drucker; Nonaka and Takeuchi), organisational management	**Organisations increase their creative capacity (the range of possible futures they can co-create with other entities) by increasing the domain and complexity of interactions they enter into (via the number and diversity of their connections).**	Autopoiesis (Maturana and Varela), complex responsive processes (Stacey), complexity theory (Kauffman)
Analytical discussion can solve organisational problems.	Reductionism, scientific realism, positivism	**When the nature of organisational conversation is creative and non-judgemental (i.e. dialogue), rather than analytic and/or defensive (i.e. discussion), the potential for insight into complex problems and transformational change in the system increases. Analysis can worsen complex problems.**	Complex responsive processes (Stacey, Shaw), single- vs. double-loop learning (Argyris and Schon), suspension (Senge, Bohm), first- vs. second-order change (Watzlawick *et al.*), Theory of Logical types (Whitehead and Russell, Bateson), group theory (Galois), chaos theory

information processing view of cognition, such as the escalation of error that can occur when analytical thinking is applied to complex problems. Rather than viewing cognition as a means of navigating or adapting to an external pre-given reality, in the complexity view individuals are seen to mutually construct themselves and their world via their cognitive interactions with one another. As such, mind is seen as contributing to the construction of reality, particularly if one takes Maturana and Varela's view that no 'things'—living or non-living—exist independently of cognition. Consciousness, language and organising are all just extensions of cognition—higher order patterns of interaction of that emerge from iterative couplings between living systems. With each new cognitive order (i.e. consciousness, language), the complexity of interactions into which the organism enters increases dramatically—enhancing its environmental responsiveness and capacity for novel behaviour (i.e. the plasticity of its behaviours). This perspective resonates with Alexander's forgotten theory of emergentism, a hierarchal process whereby spacetime—a vast unity of motion—leads to matter, matter leads to mind, and mind leads to deity (Thomas 2012; see Chapter 2).

The mainstream organisational sciences have adopted the Newtonian assumptions of reductionist science which takes its "objects of study to be composed of discrete things rather than evolving, situated relationships" (Shotter and Tsoukas 2011: 333). From this perspective, organisations are regarded as 'things', which can be manipulated and controlled by those who stand outside of them. Organisational knowledge is also regarded as a 'thing', which can be stripped from workers and commoditised into an organisational asset. However, when viewed from a complexity-based ontology, we can see that an organisation is not a thing; rather it is a stable pattern of interaction, an extension of cognition as understood from the Santiago perspective. This pattern emerges from the interactions of its members and is shaped by its couplings with other systems. Individuals, groups, organisations and society are all of the same ontological level: coherent patterns emerging from circular interactive processes (Stacey 2010). Knowledge is likewise understood as a *process* of relating, rather than a thing.

On the basis of discussions in Chapter 5, viewing organisations as *things*, rather than creative *processes*, is yet another instance of Newtonian 'flattening'—an error of logical typing whereby the emergent dimension of unified spacetime is disregarded.[34] Such errors generate paradox, such as the organisational learning paradox, whereby organisations appear to be both processes (by virtue of their behaviours) and things (by virtue of their containment of knowledge). Scholars have tried to resolve such paradoxes by ascribing complicated cognitive mechanisms involving mental models (i.e. theories-in-use), thereby keeping the Newtonian paradigm and its assumption of an external objective reality intact. As such, mental models are the metaphorical equivalent of epicycles in Ptolemy's geocentric solar system. If we replace the Newtonian paradigm with a complexity-based paradigm, and organisations are thus understood as emergent *processes*, or patterns of interaction (rather than things), the organisational learning paradox is resolved. Hence, the notion of mental models can be discarded, just as Copernicus was able to discard epicycles in his heliocentric solar system.

The phenomena of cognition, learning, knowledge and organising are key elements of knowledge transfer and research uptake. A complexity-based interpretation of these phenomena, as outlined in this chapter, can serve as an alternative framework for understanding the nature and behaviour of socio-ecological systems, and the intractable and wicked problems we encounter when we try to manage them. In Chapter 7, I will discuss the implications of this complexity-based framework in relation to science, knowledge transfer and adaptive capacity.

Notes

1 When a living system enters into a cognitive interaction, this causes changes to its internal state, but its overall circular organisation is maintained so it does not lose its identity (Maturana and Varela 1980).

2 Bateson's work differed from that of other early behavioural psychologists in that it focused on the *dyad*, the interaction involved in the behaviour, rather than on the individual behaviour itself (Visser 2007). Because Bateson saw learning as occurring in the context of relationships, his thinking is considered to some extent to be consistent with constructionism (Littlejohn and Foss 2009).

3 Capra (1997) notes that even bacteria respond to their environments, sensing chemicals, swimming toward sugars, away from acids, and moving away or toward heat or light. Such tiny acts of cognition can have profound implications as per Lovelock's Gaia hypothesis (Lovelock and Margulis 1974). Heat stimulates bacteria to grow, which increases soil weathering, which increases CO_2 removal from the atmosphere. Thus, by responding to heat, bacteria bring forth a cooler planet, albeit indirectly by coupling with a vast multitude of other processes on a multitude of scales (Capra 1997).

4 Maturana and Varela (1992) note that while the nervous system itself is both self-organising and self-referring, it is also subject to the larger circular organisation of the whole organism.

5 Maturana and Varela's autopoietic view of cognition conflicted with that of the cognitive sciences, which then viewed (and in many cases still do) the nervous system as an information processing devise that picks up information from its environment, and has inputs and outputs like that of a machine or computer. Indeed, Maturana and Varela (1992: 169) regard this information processing view of cognition as "patently wrong". According to Capra (1997), advances in cognitive sciences show that the nervous, endocrine and immune systems actually form a single cognitive network, thereby lending support Maturana and Varela's assertions.

6 The self-referential character of autopoiesis has been criticised by some as being a reflection of Maturana's radical constructivist or solipsistic epistemology (Mesjasz 2010), and his belief that we see does not exist. Maturana and Varela (1992), however, argue that the autopoietic view of cognition is neither extremely representational (concerning an objective reality) nor solipsistic/idealist (assuming there is no external, objective reality), because for the nervous system there is no 'inside' or 'outside'.

7 Total denial of representations in cognition cannot, however, account for anticipatory or 'representation-hungry' behaviour, such as reasoning about absent or non-existent circumstances, planning or imaging (Wilson and Foglia 2011).

8 Merleau-Ponty (1962) opposed ideas first proposed by Locke that perception is the causal product of atomic sensations. He proposed instead that body and perception cannot be disentangled, that perception emerges through the body's direct experience with the world.

9 Thus, cognition and perception are intrinsically temporal—they happen *in* time, rather than *over* time (Noë and Thompson 2004, citing others).

10 These developmental structures have likewise arisen from the shared co-evolutionary history between organisms; i.e. they did not evolve on their own, they emerged from relationship—out of the long-term patterns of interaction between organisms and the non-living environment.

11 Citations from interviews in the Canadian Broadcasting Corporation's *Ideas* radio programme.

12 Kauffman (2012) defines Res potentia as the realm of the ontologically real Possible, which he proposes is linked to the ontologically real Actual by quantum measurement in a new dualism. Kauffman's proposal builds on Whitehead's notion that reality is comprised of the realms of the Actual and the Possible, each of which gives rise to the other (Whitehead and Sherburne 1981).

13 The community of practice model emphasises the importance of human contact in learning. Rather than being concerned with the cognitive processes and structures involved in learning, it asks what kinds of social situations allow for learning to take place (Lave and Wenger 2001). According to Wenger (2004: 2), communities of practice are "groups of people who share a passion for something that they know how to do and who interact regularly in order to learn how to do it better". They are self-organising webs of relationships that emerge around things that matter as people share their knowledge (Wenger 1998).

14 A system caught in a *game without end* cycles through all its possible internal states without changing its overall state. This is because it cannot generate within itself the conditions or rules that would allow it change its own rules (Watzlawick *et al.* 2011).

15 The role of right hemisphere thinking is to "simultaneously 'capture' an infinite number of connections and the formation of an integral but ambiguous context due to this capture … the whole is not determined by its components, since all specific features of the whole are determined only by interconnections between these parts" (Rotenberg and Arshavsky 1991: 183).

16 Henri Poincare had his sudden insight into non-Euclidean geometry while he was boarding a bus (Ghiselin 1954: 4–5).

17 For an example, see Groucho Marx's version of the Cretan's paradox in Chapter 3.

18 Double binds are situations where the rules of a system are such that success in meeting one demand results in failure to meet another, and vice versa (Bateson 2000).

19 A classic example is China's Four Pests Campaign. Because sparrows eat grain and were seen to be 'robbing' farmers, the Chinese government rewarded citizens for killing tree sparrows. Widespread sparrow eradication resulted in one of the world's largest self-inflicted environmental disasters—unchecked by their natural predators (sparrows), plagues of locusts swept the country, destroying crops and promulgating the Great Chinese Famine and the starvation deaths of 20 million people.

20 Argyris and Schön (1996) regard these governing variables as values and assumptions in the form of 'tacit mental models'. In the complexity view, these governing variables aren't models—they're habits, or patterns, of thought and action that form via repeated coupling between and individual and the systems that surround it.

21 Taylor's scientific management involved transfer of production control from workers to management, the division of labour into simple tasks, and a move away from expensive craftsmanship to mass production, automation, and cheap unskilled labour (Khurana 2009; Waring 1991). Although widely criticised and now obsolete, scientific management had enormous influence on the manufacturing industries of the early

twentieth century; it is credited with a fifty-fold increase in productivity in industrialised countries over 100 years (Drucker 1993).

22 Logical behaviourism stemmed from Carnap's proposed application of logical positivism to psychology as a means of limiting psychological discourse to scientifically verifiable observations of behaviour, i.e. methodological behaviourism (Hauser 2005). Rather than deny the mental aspect of behaviour, Ryle and Polanyi opted for a metaphysical pluralism, whereby mind and body are seen as integrated on a tacit level (LeFave 2008; Polanyi 1969). Ryle (1949) distinguished between knowing 'how' versus knowing 'what'. Polanyi (1967) built on this with his notions of tacit and explicit and knowledge. Polanyi's premise was that we know more than we can say—for example we know how to ride a bicycle, but we cannot fully say how it is we do so. This 'know how', which we can't articulate into 'know that', is tacit knowledge. Explicit knowledge is that which can potentially be codified and therefore articulated in speech or writing, but it only has meaning when embedded in tacit knowledge (*ibid.*).

23 Drucker's (1954) management by objective (MBO) approach also involves measurement of performance against objectives, and systems of rewards and punishments. MBO became wildly popular among managers and academics in the decades that followed. However, few companies were found to actually use the method properly or in whole, and it was unclear as to how it had been successful in terms of productivity, etc. (Waring 1991). Many critics viewed Drucker's approach as a thinly disguised form of Taylor's scientific management, in that it manipulated and objectified workers into instruments for reaching a management goal (*ibid.*).

24 Stacey *et al.* (2000) note that although systems approaches were innovative in their conception of organisations as non-linear entities regulated through feedback, they erred by continuing to treat organisations as closed systems near equilibrium. As such, they failed to address the phenomena of spontaneous self-organisation—the emergent behaviour that characterises complex systems such as organisations.

25 Although he also criticises Maturana and Varela's autopoietic perspective for being a system theory and being too focused on the individual (see Stacey 2001). Examples of autopoietic interpretations of organisations include Goldspink and Kay (2010), Kickert (1993), and von Krogh and Roos (1995).

26 Although one could argue that entrenched/coherent patterns of interaction can be understood to store knowledge, which in this sense could thus be considered a 'thing'.

27 Argyris and Schön describe the counterproductive anti-learning patterns of what they call Model I organisations—organisations characterised by "defensiveness, self-fulfilling prophecies, self-fuelling processes, and escalating error" (Argyris 1982: 8).

28 Indeed, it would seem likely that organisations are more *autocatalytic* than they are *autopoietic*, meaning they lack sufficient dimensions of order and complexity (i.e. cohesiveness) needed to give them clear and substantive boundaries—individuating boundaries being a criterion for defining systems as truly autopoietic (i.e. life forms; see also Kauffman 1993 on autocatalytic systems and McMullin 2000 on the differences between autocatalysis and autopoiesis). In this sense, organisations could be viewed as large-scale analogues of hypercycles—the self-organising chemical networks believed to be the molecular precursors to life (Eigen 1978) which are *proto-conscious* (see Jordan and Ghin 2006) rather than conscious. However, Goertzel (2002), among others, believes that mounting global connectivity via the Internet may be facilitating the emergence of a higher order autonomous system (which being individuated would therefore be autopoietic) that is intelligent and conscious.

29 Although this meaning has been disputed as a misconception by some language scholars (e.g. Mair 2009).
30 As the term is applied by Seel (2006).
31 However, too much conversation, connection and diversity can also destabilise an organisation (Stacey 2001).
32 Stacey (2001) notes that corporate attempts to quickly expand organisational knowledge by merging with or acquiring another organisation often have the effect of destroying what is bought. Relationships and everyday patterns of interaction, which are the essence of organisational knowledge, cannot be bought, and typically break down as staff are removed or moved around.
33 Which themselves are actually coherent patterns of *relating* among multiple entities.
34 Regarding any phenomena as a 'thing' is the consequence of Newtonian 'flattening' of unified and emergent four-dimensional spacetime.

References

Ahrweiler, P. (2013). *Innovation in Complex Social Systems*. London: Routledge.

Allison, H. and Hobbs, R. (2006). *Science and Policy in Natural Resource Management: Understanding System Complexity*. Cambridge: Cambridge University Press.

Anderson, P. (1999). Complexity theory and organization science. *Organization Science 10*(3), 216–232.

Ardrey, R. (1970). *The Social Contract: A Personal Enquiry into the Evolutionary Sources of Order and Disorder*. New York: Atheneum.

Argyris, C. (1982). *Reasoning, Learning, and Action: Individual and Organizational*. San Francisco, CA: Jossey-Bass.

Argyris, C. (2003). A life full of learning. *Organization Studies 24*, 1178–1192.

Argyris, C. and Schön, D. A. (1974). *Theory in Practice: Increasing Professional Effectiveness*. San Francisco: Jossey-Bass.

Argyris, C. and Schön, D. A. (1978). *Organizational Learning: A Theory of Action Perspective*. Reading, MA: Addison-Wesley.

Argyris, C. and Schön, D. A. (1996). *Organizational Learning II: Theory, Methods and Practice*. Reading, MA: Addison-Wesley.

Ashby, W. R. (1952). *Design for a Brain*. New York: John Wiley.

Bahktin, M. M. (1986). *Speech Genres and Other Late Essays*. Austin, TX: University of Texas Press.

Bartunek, J. M. and Moch, M. K. (1987). First-order, second-order, and third-order change and organization development interventions: A cognitive approach. *Journal of Applied Behavioral Science 23*, 483–500.

Bateson, G. (2000). *Steps to an Ecology of Mind: Collected Essays in Anthropology, Psychiatry, Evolution, and Epistemology*. Chicago, IL: University of Chicago Press.

Baumann, O. and Siggelkow, N. (2011). Complexity and competitive advantage. In S. Macguire, P. Allen and B. McKelvey (eds), *The Sage Handbook of Complexity and Management* (pp. 494–505). London: Sage.

Beer, S. (2002). What is cybernetics? *Kybermetes 31*(2), 209–219.

Beratan, K. K. (2007). A cognition-based view of decision processes in complex social-ecological systems. *Ecology and Society 12*(1), 27.

Berkes, F. and Turner, N. J. (2006). Knowledge, learning and the evolution of conservation practices for socio ecological system resilience. *Human Ecology 34*(4), 479–494.

Bodin, A. M. (2007). *Paul Watzlawick: A Commemoration*. Palo Alto, CA: Mental Research Institute.
Bohm, D. (1994). *Thought as a System*. New York: Routledge.
Bohm, D., Factor, D. and Garrett, P. (1991). Dialogue: A proposal. Retrieved 31 August 2012 from www.david-bohm.net/dialogue/dialogue_proposal.html.
Bowden, E. M. and Jung-Beeman, M. (2003). Aha! Insight experience correlates with solution activation in the right hemisphere. *Psychonomic Bulletin and Review 10*(3), 730–737.
Broadbent, D. E. (1958). *Perception and Communication*. Oxford: Pergamon.
Brown, J. S. and Duguid, P. (1991). Organizational learning and communities-of-practice: Toward a unified view of working, learning, and innovation. *Organization Science 2*(1), 40–57.
Brown, J. S. and Duguid, P. (2000). Mysteries of the region: Knowledge dynamics in Silicon Valley. In C.-M. Lee, W. F. Miller, M. G. Hancock and H. S. Rowen (eds), *The Silicon Valley Edge: A Habitat for Innovation and Entrepreneurship* (pp. 16–45). Stanford, CA: Stanford University Press.
Buchanan, M. (2011). Power laws and the new science of complexity management. In B. McKelvey (ed.), *Complexity (Critical Concepts)* (pp. 2–10). Abingdon: Routledge.
Buckner, R. L. and Carroll, D. C. (2007). Self-projection and the brain. *Trends in Cognitive Sciences 11*(2), 49–57.
Burnes, B. (2004). Kurt Lewin and the planned approach to change: A re-appraisal. *Journal of Management studies 41*(6), 977–1002.
Burnes, B. (2005). Complexity theories and organizational change. *International Journal of Management Reviews 7*(2), 73–90.
Capra, F. (1997). *The Web of Life*. London: Flamingo.
Cave, P. (2005). Humour and paradox laid bare. *The Monist 88*(1), 135–153.
Chatwin, B. (1987). *The Songlines*. New York: Penguin Books.
Checkland, P. B. (1981). *Systems Thinking, Systems Practice*. Chichester: Wiley.
Chomsky, N. (1959). A review of B. F. Skinner's *Verbal Behaviour*. *Language 35*(1), 26–58.
Cilliers, P. (1998). *Complexity and Postmodernism: Understanding Complex Systems*. London: Routledge.
Clark, A. (1998). *Being There: Putting Brain, Body, and World Together Again*. Cambridge, MA: MIT press.
Craik, K. J. (1943). *The Nature of Explanation*. Cambridge: Cambridge University Press.
Crossan, M. and Guatto, T. (1996). Organizational learning research profile. *Journal of Organizational Change Management 9*(1), 107–112.
Cyert, R. M. and March, J. G. (1963). *A Behavioral Theory of the Firm*. Englewood Cliffs, NJ: Prentice Hall.
Davenport, T. H. and Prusak, L. (1997). *Working Knowledge: How Organizations Manage What They Know*. Boston, MA: Harvard Business School Press.
Davidow, W. H. (2011). *Overconnected: The Promise and Threat of the Internet*. New York: Delphinium Books.
Debus, A. G. (1978). *Man and Nature in the Renaissance*. Cambridge: Cambridge University Press.
Depraz, N., Varela, F. J. and Vermersch, P. (2000). The gesture of awareness: An account of its structural dynamics. In M. Velmans (ed.), *Investigating Phenomenal Consciousness* (pp. 121–136). Amsterdam: John Benjamins.
Doidge, N. (2007). *The Brain That Changes Itself: Stories of Personal Triumph from the Frontiers of Brain Science*. New York: Penguin Group.
Donaldson, A. (2005). Writing in organizational life: How a technology simultaneously

forms and is formed by human interaction. In R. D. Stacey (ed.), *Experiencing Emergence in Organizations: Local Interaction and the Emergence of Global Patterns* (pp. 142–163). New York: Routledge.

Drucker, P. (1954). *The Practice of Management*. New York: Harper & Row.

Drucker, P. (1970). *The Age of Discontinuity: Guidelines to Our Changing Society*. London: William Heinemann Ltd.

Drucker, P. (1993). *The Post-Capitalist Society*. New York: Harper Business.

Easterby-Smith, M., Antonacopoulou, E., Simm, D. and Lyles, M. (2004). Constructing contributions to organizational learning: Argyris and the next generation. *Management Learning 35*(4), 371–380.

Edmondson, A. and Moingeon, B. (1999). Learning, trust and organizational change. In M. Easterby-Smith, L. Araujo and J. Burgoyne (eds), *Organizational Learning and the Learning Organization* (pp. 157–175). London: Sage.

Eigen, M. (1978). The hypercycle: A principle of natural self□organization. *International Journal of Quantum Chemistry 14*(S5), 219–219.

Ertmer, P. A. and Newby, T. J. (1993). Behaviorism, cognitivism, constructivism: Comparing critical features from an instructional design perspective. *Performance Improvement Quarterly 6*(4), 50–72.

Festinger, L. (1957). *A Theory of Cognitive Dissonance*. Evanston, IL: Row Peterson.

Fiore, S. M. and Schooler, J. W. (1998). Right hemisphere contributions to creative problem solving: Converging evidence for divergent thinking. In M. Beeman and C. Chiarello (eds), *Right Hemisphere Language Comprehension: Perspectives from Cognitive Neuroscience* (pp. 349–371). Hillsdale, NJ: Lawrence Erlbaum.

Fischer, M. H. and Brugger, P. (2011). When digits help digits: Spatial–numerical associations point to finger counting as prime example of embodied cognition. *Frontiers in psychology 2*, 260.

Forrester, J. W. (1968). *Principles of Systems*. Waltham, MA: Pegasus Communications.

Forrester, J. W. (1975). *Collected Papers of Jay Forrester*. Portland, OR: Productivity Press.

Forrester, J. W. (1993). System dynamics and the lessons of 35 years. In K. B. Greene (ed.), *A Systems-Based Approach to Policy Making*. Dordrecht: Kluwer Academic Publishers.

Foucault, M. (2002). *Archaeology of Knowledge*. London: Routledge.

Friedman, V. J., Lipshitz, R. and Popper, M. (2005). The mystification of organizational learning. *Journal of Management Inquiry 14*(1), 19–30.

Gergen, K. J. (1997). *Realities and Relationships: Soundings in Social Construction*. Cambridge, MA: Harvard University Press.

Ghiselin, B. (1954). *The Creative Process: A Symposium*. Los Angeles, CA: University of California Press.

Goertzel, B. (2002). *Creating Internet Intelligence: Wild Computing, Distributed Digital Consciousness and the Emerging Global Brain*. New York: Kluwer Academic/Plenum Publishers.

Goldspink, C. and Kay, R. (2010). Autopoiesis and organizations: A biological view of social system change and methods for their study. *Advanced Series in Management 6*, 89–110.

Hauser, L. (2005). Behaviorism. *Internet Encyclopedia of Philosophy*. Retrieved from www.iep.utm.edu/behavior.

Heidegger, M. (1998). *Pathmarks* (ed., trans. W. McNeil). New York: Cambridge University Press.

Heylighen, F., Cilliers, P. and Gershenson, C. (2007). Complexity and philosophy. In J. Bogg and R. Geyer (eds), *Complexity, Science and Society* (pp. 117–134). New York: Radcliffe Publishing.

Holland, J. H. (1975). *Adaptation in Natural and Artificial Systems*. Ann Arbor, MI: University of Michigan Press.

Jashapara, A. (2004). *Knowledge Management: An Integral Approach*. Harlow: Pearson Financial Times.

Jones, N. A., Ross, H., Lynam, T., Perez, P. and Leitch, A. (2011). Mental models: An interdisciplinary synthesis of theory and methods. *Ecology and Society 16*(1).

Jordan, J. S. and Ghin, M. (2006). (Proto-)consciousness as a contextually emergent property of self-sustaining systems. *Mind and Matter 4*(1), 45–68.

Kauffman, S. (1993). *The Origins of Order: Self-Organization and Selection in Evolution*. New York: Oxford University Press.

Kauffman, S. (2012). Answering Descartes: Beyond Turing. In B. S. Cooper and A. Hodges (eds), *The Once and Future Turing: Computing the World*. Cambridge: Cambridge University Press.

Kennedy, P. (writer) (2011). Never in anger. Radio broadcast. *Ideas*. Canadian Broadcasting Corporation.

Kennedy, P. (writer) (2012). Imagination, parts 1 and 2. Radio broadcast. *Ideas* (producer F. Faulk). Canadian Broadcasting Corporation.

Khurana, A. (2009). *Scientific Management: A Management Idea to Reach a Mass Audience*. New Delhi: Global India Publications.

Kickert, W. J. (1993). Autopoiesis and the science of (public) administration: Essence, sense and nonsense. *Organization Studies 14*(2), 261–278.

Korzybski, A. (1958). *Science and Sanity: An Introduction to Non-Aristotelian Systems and General Semantics*. New York: Institute of General Semantics.

Kounios, J., Frymiare, J. L., Bowden, E. M., Fleck, J. I., Subramaniam, K., Parrish, T. B. *et al*. (2006). The prepared mind: Neural activity prior to problem presentation predicts subsequent solution by sudden insight. *Psychological Science 17*(10), 882–890.

Koziol, L. F., Budding, D. E. and Chidekel, D. (2012). From movement to thought: Executive function, embodied cognition, and the cerebellum. *The Cerebellum 11*(2), 505–525.

Kuhn, T. S. (1962). *The Structure of Scientific Revolutions*. Chicago, IL: University of Chicago Press.

Kuklinski, J. H. and Quirk, P. J. (2000). Reconsidering the rational public: Cognition, heuristics, and mass opinion. In A. Lupia, M. D. McCubbins and S. L. Popkin (eds), *Elements of Reason: Understanding and Expanding the Limits of Political Rationality*. Cambridge: Cambridge University Press.

Lakoff, G. and Johnson, M. (1999). *Philosophy in the Flesh*. New York: Basic books.

Laszlo, E. (1996). *Evolution: The General Theory*. New York: Hampton Press.

Laszlo, K. and Laszlo, A. (2002). Evolving knowledge for development: The role of knowledge management in a changing world. *Journal of Knowledge Management 6*(4), 400–412.

Lave, J. and Wenger, E. (2001). Legitimate peripheral participation in communities of practice. In J. Clarke, A. Hanson, R. Harrison and F. Reeve (eds), *Supporting Lifelong Learning: Perspectives on Learning and Teaching* (vol. 1, pp. 111–126). London: Routledge-Falmer/Open University.

LeFave, S. (2008). Logical behaviorism. Retrieved 20 February 2013 from http://instruct.westvalley.edu/lafave/Logical_Behaviorism.html.

Lehrer, J. (2008). The eureka hunt. *New Yorker* (28 July), 22.

Levitt, B. and March, J. G. (1988). Organizational learning. *Annual Review of Sociology 14*, 319–340.

Levy, D. (2000). Applications and limitations of complexity theory in organization theory and strategy. In J. Rabin, G. Miller and B. Hildreth (eds), *Handbook of Strategic Management* (pp. 67–86). New York: Marcel Dekker.

Lewin, K. (1946). Action research and minority problems. *Journal of Social Issues 2*(4), 34–46.

Lipshitz, R. (2000). Chic, mystique, and misconception. *Journal of Applied Behavioral Science 36*(4), 456–473.

Littlejohn, S. W. and Foss, K. A. (2009). *Encyclopedia of Communication Theory*. London: Sage Publications.

Lovelock, J. E. and Margulis, L. (1974). Atmospheric homeostasis by and for the biosphere: The Gaia hypothesis. *Tellus 26*(1–2), 2–10.

Mair, V. (2009). Danger + opportunity = crisis: How a misunderstanding about Chinese characters has led many astray. Retrieved 2 March 2013 from www.pinyin.info/chinese/crisis.html.

March, J. G. (1991). Exploration and exploitation in organizational learning. *Organization Science 2*(1), 71–87.

Maturana, H. and Varela, F. J. (1980). *Autopoiesis and Cognition: The Realization of the Living*. Dordrecht: Reidel.

Maturana, H. and Varela, F. J. (1992). *The Tree of Knowledge: the Biological Roots of Human Understanding*. Boston, MA: Shambhala.

Maxwell, N. (2000). A new conception of science. *Physics World* (August), 17–18.

McGilchrist, I. (2012). *The Master and His Emissary: The Divided Brain and the Making of the Western World*. New Haven, CT: Yale University Press.

McMullin, B. (2000). Remarks on autocatalysis and autopoiesis. *Annals of the New York Academy of Sciences 901*(1), 163–174.

Mead, G. H. (1934). *Mind, Self and Society*. Chicago, IL: Chicago University Press.

Meadows, D. H. (1999). *Leverage Points: Places to Intervene in a System*. Hartland, VT: Sustainability Institute.

Merleau-Ponty, M. (1962). *Phenomenology of Perception* (trans. C. Smith). London: Routledge & Kegan Paul. (French edition: *Phénoménologie de la perception*, Paris: Gallimard, 1945).

Mesjasz, C. (2010). Complexity of social systems. *Acta Physica Polonica A 117*(4), 707–715.

Neisser, U. (1967). *Cognitive Psychology*. New York: Appleton-Century-Crofts.

Nisbett, R. E. and Wilson, T. D. (1977). Telling more than we can know: Verbal reports on mental processes. *Psychological Review 84*(3), 231–259.

Noë, A. and Thompson, E. (2004). Are there neural correlates of consciousness? *Journal of Consciousness Studies 11*(1), 3–28.

Nonaka, I. and Takeuchi, H. (1995). *The Knowledge Creating Company: How Japanese Companies Create the Dynamics of Innovation*. New York: Oxford University Press.

Nyhan, B. and Reifler, J. (2010). When corrections fail: The persistence of political misperceptions. *Political Behavior 32*(2), 303–330.

Nystrom, P. C. and Starbuck, W. H. (1984). To avoid organizational crisis, unlearn. *Organizational Dynamics 12*(4), 53–65.

Ong, W. J. (2002). *Orality and Literacy*. New York: Taylor & Francis.

Park, P. (1999). People, knowledge, and change in participatory research. *Management Learning 30*(2), 141–157.

Polanyi, M. (1967). *The Tacit Dimension*. New York: Doubleday.

Polanyi, M. (1969). *Knowing and Being: Essays* (ed. M. Grene). Chicago, IL: University of Chicago Press.

Popper, M. and Lipshitz, R. (1998). Organizational learning mechanisms. *Journal of Applied Behavioral Science 34*(2), 161–179.
Prigogine, I. (1967). Dissipative structures in chemical systems. In S. Claesson (ed.), *Fast Reactions and Primary Processes in Chemical Kinetics*. New York: Interscience.
Prigogine, I. (1997). *The End of Certainty*. New York: Free Press.
Proulx, J. (2008). Some differences between Maturana and Varela's theory of cognition and constructivism. *Complicity: An International Journal of Complexity and Education 5*(1), 11–26.
Putnam, H. (ed.). (1960). *Minds and Machines*. New York: New York University Press.
Richardson, K. A. (2004). Systems theory and complexity: Part I. *E:CO 6*(3), 75–79.
Richardson, K. A. (2005). The hegemony of the physical sciences: An exploration in complexity thinking. *Futures 37*(7), 615–653.
Rotenberg, V. and Arshavsky, V. (1991). Psychophysiology of hemispheric asymmetry. *Integrative Physiological and Behavioral Science 26*(3), 183–188.
Ryle, G. (1949). *The Concept of Mind* London: Hutchinson.
Sandkühler, S. and Bhattacharya, J. (2008). Deconstructing insight: EEG correlates of insightful problem solving. *PLoS One 3*(1), e1459.
Sartre, J.-P. (1960). *The Transcendence of the Ego: An Existentialist Theory of Consciousness* (trans. F. Williams and R. Kirkpatrick). New York: Hill & Wang.
Schacter, D. L. and Addis, D. R. (2000). The cognitive neuroscience of constructive memory: Remembering the past and imagining the future. *Philosophical Transactions of the Royal Society B 362*, 773–786.
Schooler, J. W., Ohlsson, S. and Brooks, K. (1993). Thoughts beyond words: When language overshadows insight. *Journal of Experimental Psychology: General 122*(2), 166.
Schutt, P. (2003). The post-Nonaka knowledge management. *Journal of Universal Computer Science 9*(6), 451–462.
Schwartz, J. and Beyette, B. (1996). *Brain Lock: Free Yourself from Obsessive-Compulsive Behavior: A Four-Step Self-Treatment Method to Change Your Brain Chemistry*. New York: Regan Books.
Seel, R. (2006). Emergence in organisations. Retrieved 3 September 2010 from www.new-paradigm.co.uk/emergence-2.htm.
Senge, P. M. (1990). *The Fifth Discipline: The Art and Practice of the Learning Organization*. New York: Currency Doubleday.
Senge, P. M., Scharmer, C. O., Jaworski, J. and Flowers, B. S. (2004). *Presence: Human Purpose and the Field of the Future*. New York: Doubleday.
Shackley, S., Wynne, B. and Waterton, C. (1996). Imagine complexity: The past, present and future potential of complex thinking. *Futures 28*(3), 201–225.
Shammi, P. and Stuss, D. T. (1999). Humour appreciation: A role of the right frontal lobe. *Brain 122*(4), 657–666.
Shaw, P. (2002). *Changing Conversations in Organisations: A Complexity Approach to Change*. London: Routledge.
Shotter, J. (1994). *Conversational Realities: Constructing Life through Language*. London: Sage Publications.
Shotter, J. and Tsoukas, H. (2011). Complex thought, simple talk: An ecological approach to language-based change in organizations. In S. Macguire, P. Allen and B. McKelvey (eds), *The Sage Handbook of Complexity and Management* (pp. 333–348). London: Sage.
Sides, J. and Citrin, J. (2007). How large the huddled masses? The causes and consequences of public misperceptions about immigrant populations. Paper presented at the Annual Meeting of the Midwest Political Science Association, Chicago, 12–15 April.

Smith, M. (2001). Peter Senge and the learning organisation. *Encyclopedia of Informal Education*. Retrieved from www.infed.org/thinkers/senge.htm.

Snowden, D. (2003). Complex acts of knowing: Paradox and descriptive self-awareness. *Bulletin of the American Society for Information Science and Technology 29*(4), 23–28.

Stacey, R. (1995). The science of complexity: An alternative perspective for strategic change processes. *Strategic Management Journal 16*(6), 477–495.

Stacey, R. (1996a). *Complexity and Creativity in Organizations*. San Francisco, CA: Berrett-Koehler.

Stacey, R. (1996b). Emerging strategies for a chaotic environment. *Long Range Planning 29*(2), 182–189.

Stacey, R. (2001). *Complex Responsive Processes in Organizations*. London: Routledge.

Stacey, R. (2003). *Strategic Management and Organisational Dynamics: The Challenge of Complexity*. Harlow: Prentice-Hall.

Stacey, R. (2005). Local and global processes in organizational life. In R. D. Stacey (ed.), *Experiencing Emergence in Organizations: Local Interaction and the Emergence of Global Pattern.* (pp. 17–47). New York: Routledge.

Stacey, R. (2010). The emergence of knowledge in organizations. In A. Tait and K. A. Richardson (eds), *Complexity and Knowledge Management: Understanding the Role of Knowledge in the Management of Social Networks* (pp. 41–56). Charlotte, NC: Information Age Publishing.

Stacey, R., Griffin, D. and Shaw, P. (2000). *Complexity and Management: Fad or Radical Challenge.* New York: Routledge.

Stark, B. (2006). The global political system: A dynamical system within the chaotic phase. Paper presented at the International Conference on Complex Systems, 25–30 June, Boston, MA.

Tan, J.-S. (2013). Trends in organizational change. *Reference for Business: Encyclopedia of Business* (2nd edition). Retrieved from www.referenceforbusiness.com/management/Tr-Z/Trends-in-Organizational-Change.html.

Tate, W. (2005). Working with the shadow side of organisations. *Developing HR Strategy* (May), 22–25.

Taylor, F. W. (1911). *The Principles of Scientific Management*. New York: Harper & Brothers.

Thelen, E. and Smith, L. B. (1996). *A Dynamic Systems Approach to the Development of Cognition and Action*. Cambridge, MA: MIT Press.

Thomas, E. (2012). Samuel Alexander (1859–1938). *Internet Encyclopedia of Philosophy*. Retrieved from www.iep.utm.edu/alex-sam.

Tolle, E. (2006). *A New Earth: Awakening Your Life's Purpose*. New York: Penguin.

Varela, F., Thompson, E. T. and Rosch, E. (1991). *The Embodied Mind: Cognitive Science and Human Experience*. Cambridge, MA: MIT Press.

Varela, J. A. (1971). *Psychological Solutions to Social Problems*. New York: Academic Press.

Visser, M. (2007). Deutero-learning in organizations: A review and a reformulation. *Academy of Management Review 32*(2), 659–667.

Von Krogh, G. and Roos, J. (1995). *Organizational Epistemology*. Basingstoke: Palgrave Macmillan.

Vygotsky, L. S. (1962). *Thought and Language*. Cambridge, MA: MIT Press.

Waring, S. P. (1991). *Taylorism Transformed: Scientific Management Theory Since 1945?* Chapel Hill, NC: University of North Carolina Press.

Watzlawick, P., Beavin, J. H. and Jackson, D. D. (1967). *Pragmatics of Human Communication: A Study of Interactional Patterns, Pathologies and Paradoxes*. New York: W. W. Norton.

Watzlawick, P., Weakland, J. H. and Fisch, R. (2011). *Change: Principles of Problem Formulation and Problem Resolution*. New York: W. W. Norton.

Weick, K. E. (1979). *The social Psychology of Organizing*. Reading, MA: Addison-Wesley.

Wenger, E. (1998). *Communities of Practice: Learning, Meaning, and Identity*: New York: Cambridge University Press.

Wenger, E. (2004). Knowledge management as a doughnut: Shaping your knowledge strategy through communities of practice. *Ivey Business Journal 68*(3), 1–8.

Whitehead, A. N. and Russell, B. (1913). *Principia Mathematica*. Cambridge: Cambridge University Press.

Whitehead, A. N. and Sherburne, D. W. (1981). *A Key to Whitehead's Process and Reality*. Chicago, IL: University of Chicago Press.

Wilson, A. D. and Golonka, S. (2013). Embodied cognition is not what you think it is. *Frontiers in Psychology 4*. Retrieved from http://journal.frontiersin.org/article/10.3389/fpsyg.2013.00058/full.

Wilson, M. (2002). Six views of embodied cognition. *Psychonomic Bulletin and Review 9*(4), 625–636.

Wilson, R. A. and Foglia, L. (2011). Embodied cognition. *Stanford Encyclopedia of Philosophy* (Fall 2011 edition). Retrieved from http://plato.stanford.edu/entries/embodied-cognition.

Wittgenstein, L. (1980). *Culture and Value* (trans. P. Winch). Oxford: Blackwell.

Wittgenstein, L. (2010). *Philosophical Investigations* (trans. P. M. S. Hacker and J. Schulte). Oxford: Wiley-Blackwell.

Wuthrich, C. (2010). Emergent spacetime in quantum theories of gravity. Retrieved 20 September 2012 from http://philosophyfaculty.ucsd.edu/faculty/wuthrich/research/emergentspacetime.html.

Yuan, Y. and McKelvey, B. (2004). Situated learning theory: Adding rate and complexity effects via Kauffman's NK model. *Nonlinear Dynamics, Psychology and Life Sciences 8*(1), 65–10.

7 Implications

Rethinking science, knowledge transfer and adaptive management

A picture held us captive. And we could not get outside it, for it lay in our language and language seemed to repeat it to us inexorably.

(Ludwig Wittgenstein, *Philosophical Investigations*)

Puzzles, paradigms and paradox

The next step in resolving intractable problems requires reframing the problem at hand in accordance with our revised set of assumptions (i.e. within a new paradigm). This brings us to the work of Thomas Kuhn. Kuhn started his career as a physicist, but shifted his focus to the history and philosophy of science in the 1950s. In 1962 he published *The Structure of Scientific Revolutions*. In his book, Kuhn traces the history of science and concludes that it progresses through periods of slow stable growth, punctuated by revolutionary paradigm shifts marked by crisis before one prevailing theory is discarded and replaced by another. He rejects the idea that science precedes in a gradual progression that incrementally reveals the true nature of reality. Rather, his historical view suggests that as a new disorganised science becomes more and more organised, it eventually structures into a 'normal science' consisting of a single paradigm. According to Kuhn, the stable, 'normal' periods of scientific change are *qualitatively* different from the changes that take place during paradigm shifts (Bird 2011). Chalmers (2002) describes a scientific paradigm as comprising the beliefs, general theoretical assumptions, laws and techniques adopted by its particular scientific community (e.g. the Newtonian paradigm). However, as discussed earlier, Gödel's incompleteness theorem shows that there are problems that cannot be solved within the paradigm they are situated. This is particularly so in the case of intractable or 'wicked' problems.

In normal science, effort is focused on validating and reinforcing the existing paradigm, which Kuhn describes as puzzle-solving governed by the rules and premises of the paradigm. Normal scientists therefore assume "that a paradigm provides the means for the solution of the puzzles posed within it", and are therefore uncritical of it (Chalmers 2002: 110). As such, observations or theories that fail to fit the dominant paradigm are often discredited or dismissed as anomalies by the normal scientific community. However, as challenging evidence piles up and can no longer be ignored, a point of crisis is reached. This is resolved when a new paradigm develops, to which scientists increasingly defect, until the old paradigm is finally abandoned in favour of the new (*ibid.*). Hence, Kuhn challenged long-held empirical and positivist assumptions of an objective reality which is methodically revealed through the workings of science. His work shows that scientific reality is largely a social construction, susceptible to social and cultural influences. This evidenced by the divergent reactions to Darwin's theory of evolution in England and Russia described in Chapter 2. Todes (2009) states:

> Researchers bring their life experiences and culture with them into the field and laboratory, and in the course of their investigations actively originate, interpret, develop and reject metaphorical pathways. As is shown by the reception of Darwin's theory in Russia, the deployment and criticism of metaphors are part of the ineffably human process by which scientists mobilize their experiences and values to explore the infinite complexity of nature.
>
> (Todes 2009: 37)

The discussions outlined in Chapter 5 indicate how the limits of reductionist science are being made apparent by Gödel's incompleteness theorem, quantum uncertainty and the unpredictability of chaotic systems. Heylighen *et al.* (2007) suggest that since we cannot completely describe complex systems, we cannot devise unchanging and non-provisional sets of rules to control their behaviour. Richardson (2005, 2008) likewise states that the emerging complexity sciences are forcing us to revisit the nature of scientific knowledge and the limits of scientific methods. He notes that while many of the criticisms of science have been made by non-scientists, complexity science is now leading a critique of science from within its own paradigm, ironically enabled by dogmatically applied reductionist science.

However, just as Kuhn predicts, many complexity researchers (i.e. 'normal' scientists) are still implicitly clinging to the Newtonian paradigm, trying to find mathematical laws of complexity that will allow the restoration of some form order and determinism to the unpredictable, non-linear world they are trying to understand (Heylighen *et al.* 2007; Tsoukas and Hatch 2001). This is seen in the work of Holland (1987), who decomposes the workings of complex adaptive systems as if they were information-processing machines, with the aim of improving prediction and control (Chapter 4). Researchers such as Holland appear to have gained 'representational' knowledge of complex systems, but without having embodied this knowledge tacitly. Thus their assumptions about the fundamental nature of reality remain intact, and the Newtonian paradigm unchallenged. Complexity thinking without a shift in ontology simply becomes an extension of systems thinking, a more sophisticated means of trying to predict and control system behaviour.

Heylighen *et al.* (2007) suggest this is because complexity researchers come from a relatively recent scientific background and have not yet reflected about the philosophical foundations of their approaches. This is a legacy of the pre-Second

World War divorce logical positivists precipitated between science and philosophy, described in Chapter 2. This same divorce led to the abandonment of ecology's idealist and organicist roots after Tansley (1935) declared war on organicism in 1935, replacing early ecology's focus on pattern, holism and organising relations with his materialist approach and a trophic dynamics focus. Quantum physicist Carlo Rovelli (2012) laments that scientists of today are seldom educated in philosophy, and no longer reflect on the metaphysical implications and assumptions of their work.[1]

However science does not just seek to manipulate nature, it also seeks to understand, to "dig deeper into questions that have been asked generation after generation" (Prigogine and Stengers 1984: 291). The works of other complexity researchers, such as Kauffman and Prigogine, show fundamental ontological and epistemological shifts that challenge the reductionist and determinist assumptions of the Newtonian paradigm. Complexity thinking in their cases has resulted in a new understanding of reality, that of an immaterial, indeterminate, and interconnected universe being perpetually created, as outlined in Chapter 5. This is a radical departure from the Newtonian view of the universe as a shifting atomic configuration of matter and energy. In this final chapter, I will examine how this departure exposes the limits of reductionist science, and the implications for knowledge transfer and adaptive capacity in environmental management.

Limits of reductionist science

Stability

Richardson (2005) argues against the Newtonian possibility of complete and certain and knowledge, stating that "Contrary to popular belief, science is not capable of considering all phenomena. In fact, it is quite inflexible in its requirements" (*ibid.*: 617). In the Newtonian paradigm entities or phenomena must be *stable* to be assumed to be 'real'. In being stable, their 'pattern' can be distinguished from the 'background' (i.e. the larger web of interactions of which they are part). Indeed, the main requirement of scientific explanation is that the phenomena of interest *must be very stable*—sufficiently so that they can be repeatedly measured and examined, and their behaviours predicted via the principles of cause and effect (*ibid.*).

According to Richardson (*ibid.*), the scientific method yields excellent results when applied to such situations, especially when dealing with mechanical 'things' that are constructed of parts that stay the same. He notes, however, that when dealing with complex self-organising systems, which are constantly evolving and their boundaries continuously changing and emerging—stability and predictability begin to break down, particularly at the scale of social systems. Here the limits of the scientific method become evident. Snowden (2003, citing others) states that this is because cause-and-effect are intimately intertwined and inseparable in complex systems. He illustrates the vital difference in dealing with a *complex* versus a *complicated* system as follows:

> When a rumour of reorganisation surfaces: the complex human system starts to mutate and change in unknowable ways; new patterns form in anticipation of the event. On the other hand, if you walk up to an aircraft with a box of tools in your hand, nothing changes.
>
> (Snowden 2003: 25)

It should be reiterated, however, that from the perspective of the complexity-based ontology outlined in Chapter 5, stability is scale and observer dependent. Systems become more or less chaotic depending on the spatial and temporal scales at which they are observed. A phenomenon that appears unstable at one scale may be seen as ordered and stable at another. Thus, the applicability of the scientific method is likewise scale dependent—*it is not universal.* For example, reductionist science may provide good explanatory capability at the cellular level (cells also being complex self-organising systems), but its predictive powers diminish as you move up in scale and dimensions of complexity, from cell to organism, to ecosystem, then to biosphere. As Bateson states:

> Not only can we not predict into the next instant of the future, but, more profoundly, we cannot predict into the next dimension of the microscopic, the astronomically distant, or the geologically ancient. As a method of perception—and that is all science can claim to be—science, like all other methods of perception, is limited in its ability to collect the outward and visible signs of whatever may be truth. … Science probes, it does not prove.
>
> (Bateson 2002: 27)

To apply reductionist science to complex self-organising systems that are not highly stable and coherent, we essentially have to fake stability (Richardson 2005), as did engineers of the last century with their linear approximations (Capra 1997). And in so doing, we shore up our beliefs and perceptions that these dynamic, unpredictable systems can be tamed and controlled, if only we gather sufficient information about them. Thus the Newtonian scientific paradigm is perpetuated, and evidence pointing to alternate ways of understanding the world is denied or rejected, as predicted by Thomas Kuhn.

Emergence

The other limitation of the scientific method is in dealing with emergent phenomena. Mainstream reductionist science concerns itself with parts and quantities. It presumes that all 'real' phenomena and the relationships between them can be somehow quantified, and their behaviours predicted as per the laws of cause-and-effect. According to the complexity-based paradigm outlined in Chapter 5, however, order and chaos are spatially and temporally laminated in the dynamic, nested patterns of interaction (i.e. systems) that comprise an emergent, self-organising reality. The chaotic laminations are transition zones—dynamic, observer-dependent boundaries that signal a qualitative change in pattern, a jump to a different system level. The premises, rules and descriptors that apply to one level of order cannot be applied to a higher level of emergent order. A new framework and set of governing rules and premises is required (i.e. the whole is greater than the sum of its parts), as per Gödel's incompleteness theorem and the theory of logical types. As the examples presented in Chapter 3 demonstrate, attempting to apply logical, cause-and-effect analysis *across system levels* amounts to comparing apples to oranges, and in so doing generates *paradox*. Bateson invites his readers to this "psychological experience" by which he demonstrates the "frailty of the human computer" (Bateson 2000: 463), and how confusion multiplies when attempting to define the difference between differences between logical levels (i.e. meta-difference), while holding our Newtonian 'maps' depicting a quantified territory:

> First note that differences in texture are *different* (a) from differences in colour. Now note that differences in size are *different* (b) from differences in shape. Similarly ratios are *different* (c) from subtractive differences. Now let me invite you, as disciples of Korzybski, to define the differences between 'different (a),' 'different (b),' and 'different (c)' in the above paragraph. The computer in the human head boggles at the task.
>
> (Bateson 2000: 463–464)

Prediction

Thus, while reductionist science can deploy different sets of tools to effectively quantify and analyse phenomena (provided they are sufficiently stable relative to the observer) at different logical levels, it cannot explain the transition, or relations *between* the logical levels. This is because *it cannot quantify across levels*—the rules and premises governing each level are *qualitatively different*, and so too must be the metrics that are deployed. Because no singular set of assumptions, rules and metrics and can cross them, chaotic zones embed the larger system whole with uncertainty, thereby ruling out cause-and-effect prediction[2] across system boundaries, and making for irreducible uncertainty. This serves to intrinsically limit the explanatory power of any given quantitative method[3] of description. Thus, the assumption that uncovering more rules and premises will lead to a greater ability to describe the world is subject to diminishing returns.[4] Efforts to explain a situation "through development and testing of hypotheses derived from a theory will converge towards some maximum point" (James 2002: 217)—an upper limit of description, prediction and explanation.

In sum, reductionist methods have no means of bridging logical/system levels, because levels are separated by *qualitative* not quantitative divides. This relates to neuroscience and psychological research, detailed in Chapter 6, indicating that insight into solving complex problems occurs spontaneously, in the *absence* of analytic thinking (Kounios *et al.* 2006; Watzlawick *et al.* 2011). Bridging qualitative divides involves activating the pattern recognition mode of the brain's right hemisphere, from which our intuition, creativity, artistry and aesthetic sensitivities arise (Bateson 2000; McGilchrist 2012). Notably, these are modes of thought not conventionally valued or applied within the mainstream scientific paradigm. Also notable are Einstein's repeated exhortations that he learned more from Fyodor Dostoyevsky than he did from any physicist (Prigogine 1997). We cannot privilege the quantitative over the qualitative.

Newtonian paradigm and the role of science

Complexity thinking helps us recognise that, while very useful when dealing with stable phenomena, the scientific method, as applied by 'normal' science,[5] is *not* universally applicable. Attempting to apply system metrics (and conjoined rules and premises) across system levels in order to quantify them is a denial of complexity. When insisting that reductionist methods can and should be applied to all 'real' phenomena, scientists are signalling their allegiance to the mainstream scientific paradigm. Not only does this reinforce the Newtonian paradigm of a predictable, deterministic universe, it is also an error in logical typing. Such errors generate undesirable paradoxical situations from which we have difficulty escaping, despite our best efforts. These include vicious circles, games without end and deadlocks—referred to in planning and environmental management as *wicked problems*. The following sections explore how the concepts of knowledge transfer and adaptive capacity stem from a Newtonian understanding of the role of science, and the paradoxical errors this creates.

Science and the project of modernity

According to Heylighen *et al.* (2007), the role of science in the Newtonian paradigm is to improve the 'map' in the observers' minds, so it is a more accurate reflection of reality. This achieved through empirical and objective observation of the external world, using the scientific method to collect and analyse information about it, thereby further completing the 'picture' of external reality. Ultimately this will lead to a perfect picture, or map, of reality, enabling us to make accurate predictions of all phenomena. Heylighen *et al.* (*ibid.*) note that this view requires making precise distinctions between the different components, properties and states of the systems under observation (i.e. reductionism). It also assumes that these distinctions will be the same for all observers (i.e. they will be objective). The authors thus argue that mainstream science is not a creative enterprise, it is one of *uncovering*. On this basis, positivist and realist philosophies of science see science as progressively revealing the 'true' nature of reality. Table 7.1 provides a summary comparison of the role and application of science from a Newtonian perspective versus the complexity-based perspectives outlined in Chapters 5 and 6.

Heylighen *et al.* argue that Newtonian thinking that favours determinism over free will is what ultimately led to the dominance of 'rationality' in social and economic modelling:

Table 7.1 A comparison of the role and application of science from the perspective of the mainstream Newtonian paradigm versus the proposed complexity-based paradigm

Mainstream Newtonian paradigm	*Sources*	*Proposed complexity-based paradigm*	*Sources*
Reductionist methods can be universally applied to all phenomena.	Determinism, reductionism, positivism	**Reductionist methods are ineffective at describing/predicting unstable or emergent phenomena.**	Chaos theory, complexity theory
All phenomena can be quantified and accurately predicted, given sufficient information.	Determinism, reductionism, positivism	**Quantitative methods cannot be applied across emergent/chaotic boundaries, ruling out cause-and-effect prediction. Qualitative methods are required.**	Chaos theory, complexity theory
Given complete information humans make rational decisions.	Rationale choice theory, game theory	**Humans make decisions based on constraints, opportunities and power relations imposed by relationships.**	Autopoiesis (Maturana and Varela), complex responsive processes (Stacey), social constructivism (Bakhtin), postmodernism (Foucault)
Science progresses by gradually revealing the true nature of reality.	Positivism, scientific realism	**Science normally works to validate existing paradigms until crisis causes revolutionary shifts to new ones.**	Paradigmatic revolutions (Kuhn)

> Assuming perfect information about the utility of possible options, the actions of mind then become as determined or predictable as the movements of matter. This allowed social scientists to describe human agency with most of the Newtonian principles intact.
>
> (Heylighen *et al.* 2007: 121)

According to the authors, this has led to a widespread assumption that, given complete information, people will make 'rational choices', meaning people will predictably choose options with the greatest 'utility', which represents some objective value or 'goodness'. They argue that this has led to an assumption that increases in scientific knowledge typically lead to increases in global utility or well-being (i.e. linear progress). This belief in the incremental uncovering of the 'true' nature of reality by science, combined with notions of rational choice and linear progress, defines the Western project of *modernity* (*ibid.*), and how we believe the world works. As a consequence, our mainstream view of the role science in society is fundamentally premised on a string of materialist assumptions:

1. Reality exists outside of and independently to us (*materialist ontology*).
2. Reality it is knowable (*materialist epistemology*).
3. Knowledge is created as we observe reality through our senses (*empiricism*).
4. We use this knowledge to create mental models of reality (*cognitivism*).
5. Our knowledge and mental models are imperfect and incomplete (*scientific realism*).
6. Science works to progressively uncover the true nature of an objective reality (*scientific realism/positivism*).
7. Humans process information like computers, and make rational decisions based on their knowledge and mental models of reality (*rational choice theory/cognitivism*).
8. When given complete information about the utility of different options, humans will always choose the 'best' one (*rational choice theory/determinism*).
9. The more information we have about the true nature of reality, the better choices we'll make, and the greater our wellbeing (*global utility/linear progression*).
10. Therefore scientific knowledge leads toward perfect knowledge/models of an objective reality, which leads to perfect choices, which leads to an ideal state of well-being.
11. Therefore increases in scientific knowledge = increase in our well-being.

Thus, in the Newtonian paradigm, complete, objective and certain knowledge of past and future reality is possible. Uncovering this reality then becomes the mission of science, one which will ultimately increase our global well-being. From this string of assumptions emerges the rationale for *knowledge transfer*.

Rethinking knowledge transfer

Whereas knowledge management is concerned with managing knowledge as an organisational asset, knowledge transfer is concerned with transferring knowledge from one organisation to another, typically from a research body to a management agency, or from researchers to practitioners. As outlined in Chapter 1, the objective of knowledge transfer is to see scientific information applied in practice and decision-making; knowledge transfer without adoption is considered a failure (Roux *et al.* 2006).

Like knowledge management, knowledge transfer, as its name suggests, is premised on Newtonian assumptions of knowledge as a 'thing', and the computer or information processing view of cognition and learning. It is also premised on the assumptions of rational choice theory.[6] According to Hausman (2008), rational choice theory attempts to explain and predict socio-economic patterns and phenomena arising from multiple individual choices, as individuals attempt to maximise their benefits and minimise their costs. These choices are assumed, to some extent, to be rational. The theory is predicated on the notion that when given an option, an individual (i.e. agent in a system) will always chose the option with the greatest *utility*; utility being an *objective* measure of value or 'goodness'. Game theory and social choice theory are derivatives of rational choice theory. In the 1950s and 1960s, the development of complex algorithms predicting individual choices facilitated widespread uptake of rational choice theory as a 'universal theory of human behaviour'. The theory was taken up by American social scientists committed to using scientific approaches, as well as philosophers, mathematicians, operations researchers and computer scientists (Herfeld 2012). As such, rational choice became an "article of faith" among mainstream economists until the 1990s (Hodgson 2012). It is also the foundation of neo-classical economics.

However, since the 1990s, the assumptions and predictive capacity of rational choice models[7] have been widely disputed (including by six Nobel Prize winners for economics), particularly the notion that people consistently act in ways that maximise their rewards (*ibid.*). Scholars of behavioural economics have since recognised that people are driven by a much more complex suite of factors than that presented by rational choice theory. Behavioural economist Daniel Read points out that people:

> ignore important decision factors, put undue weight on some factors relative to others, plan to do the right thing but fail to follow through with those plans, they are more sure about their decisions or beliefs than they should be, they trust others more than they should, and they even fail to do simple calculations that could solve important problems.
>
> (Read cited in Foley and Griffiths 2011: 21)

As the assumption of human rationality came under increasing fire in mainstream economics during the 1990s, rational choice theory was meanwhile gaining popularity in sociology, particularly with the rise of game theory (Hodgson 2012). Despite its critics, the 'rationality' of individual behaviour remains the unquestioned base assumption of contemporary economic analysis and is the "touchstone by which mainstream economists recognize each other" (Foley 2004: 329). Policy-making also continues to be largely premised on cost-benefit models of human decision-making derived from rational choice theory (Foley and Griffiths 2011).

This is exemplified by McNie (2007), who argues that the main goal of knowledge transfer is to create useful information for decision-makers; and by

Roux *et al.* (2006), who state that its objective is the unobstructed flow of knowledge between managers and researchers. In both cases, the authors regard knowledge as a 'thing', with the goal being exchange of information that will lead to better decision-making. Indeed, much of the knowledge transfer literature is focused on ways to make information more palatable and relevant to managers and policy makers, such as by improving how it is packaged, and increasing its salience, credibility and legitimacy (McNie 2007, citing others therein).

However, as touched on in Chapter 1, there is also an increasing recognition in the literature that research uptake is improved if knowledge transfer is accompanied by some form of relationship building and participation (see Chapman *et al.* submitted). For example, McNie (2007) refers to the importance of social capital in ensuring the legitimacy of scientific information, and recommends strengthening links between scientists and policy-makers through participatory processes, adaptive management, and boundary organisations/spanners. Roux *et al.* (2006) emphasise the importance of the 'tacit' dimension of knowledge and the need to create common spaces where managers and scientists can meet and share knowledge. They likewise propose that 'unified learning systems' are needed, whereby scientists and decision-makers work together to create new knowledge, as do numerous other authors (e.g. Born *et al.* 2009; Cash and Moser 2000; Hunt and Shackley 1999; Kainer *et al.* 2009; Lemos and Morehouse 2005; Shackleton *et al.* 2009). This pertains to the work of Park (1999), who identifies three types of knowledge that can be produced during participatory research and which evoke different levels of behavioural change. Representational (or explicit) knowledge consists of facts and figures, and seldom results in practice-change on its own. Relational and reflective knowledge, which together comprise tacit knowledge, are generated as scientists and stakeholders build trust and relationships, and begin to reflect on their underlying assumptions. From relational and reflective knowledge emerges people's commitment to act, and thereby a greater likelihood that representational knowledge will be applied in practice.

Clearly, these authors have found that decisions and policies are more likely to incorporate scientific perspectives if relationships have somehow been established between scientists and decision-makers. However, the strength of the information processing and rational choice paradigms is such that the authors presume this is so because the relationships made for more salient, credible and legitimate information, which in turn led to science being incorporated into decision making, rather than assuming science uptake was simply the result of the relationships themselves.

Just as Ptolemy had to conjure 'epicycles' to make his geocentric version of the universe work, so to have knowledge transfer scholars had to conjure 'mental models' to make their Newtonian mind-as-a-computer version of cognitive and social systems work. In both situations, the addition of an 'extra step' to reconcile observations with the paradigm leads to theories that are unnecessarily convoluted and which generate paradox. The repercussions of these convolutions are addressed in the discussion that follows.

Implications for environmental management

> The management of complex natural systems as if they were simple scientific exercises has brought us to our present mixture of triumph and peril.
>
> (Funtowicz and Ravetz 2003)

In her knowledge transfer review,[8] McNie (2007) states that we have little understanding of decision processes and lack a framework for understanding them. She also suggests that the problem of research uptake can be resolved by improving the usefulness of scientific information for decision-making. The arguments outlined in Chapter 6, however, suggest that it is *relationships* and patterns of interaction—not information processing—that directly drive most decisions and system behaviour. 'Knowledge' of a system is better understood as a function of its internal and external connectivity (i.e. relationships between its agents, which confer the system with *plasticity*), than as the amount of information held inside the heads of its individual members. Park's (1999) notion of relational knowledge—despite his treating it as a 'thing' rather than a process of interacting—captures this. It is likewise reflected in Folke's description of nodes in human networks as reservoirs of collective knowledge (Folke 2003; Folke *et al.* 2005).

The discussions in Chapter 6 also suggest that analytic application of scientific information is a form of single-loop learning (Argyris and Schön 1996), which, in the case of intractable or wicked problems, often exacerbates the issue under question. Cognitive dissonance theory similarly indicates that providing people with information that contradicts their existing beliefs often causes them to become hostile and entrench their existing behaviours, as per cognitive backfire (Nyhan and Reifler 2010). This is a particularly relevant in arenas where anti-science attitudes have been forged by past resource-use planning conflicts.

Hence, scientists are more likely to influence a system's behaviours via establishing relationships and actively connecting with other components of the system, than they are by passively supplying information to others. This gives credence to Lomas's (2007: 130) earlier noted contention that "human interaction [is] the engine that drives research into practice". The central role of relationships versus information in decision-making explains why the documented barriers to knowledge transfer are largely linked to culture clashes that impede interaction between scientists and managers (Roux *et al.* 2006). It is also why there is increasing focus on collaborative approaches to conducting research, such as research for development programs, participatory research, communities of practice, knowledge interfacing and extension. In addition, it explains why impersonal forms of mostly one-way information exchange—such as newsletters, factsheets, and meetings and presentations—are largely ineffective, and why education or outreach efforts that provide people with information on a particular issue often have little or no impact on behaviour (Beratan 2007, citing others therein). It also pertains to increasing recognition of the value of using knowledge brokers to catalyse relationships between researchers and practitioners. This is illustrated by Michaels (2009), who shows how the relationship building aspect knowledge brokering should increase with intended strength of action. From a complexity perspective, social learning[9]—as it has been applied in collaborative planning[10] and adaptive co-management[11]—appears to supply the best option for successful research uptake. Social learning is action-oriented, and integrates learning with decision-making via collaborative processes where scientists are part of the decision-making fora.

From the discussions above and in Chapter 6 we can now see how assumptions of the Newtonian paradigm create a paradoxical understanding of knowledge transfer. Knowledge is not an object or thing that can be transferred or exchanged, it is a process of relating. In assuming it is a 'thing' one commits an error in logical typing. Supplying people with credible scientific information has limited impact when people seldom make decisions rationally or consciously apply logic to their thinking. When decisions are made that incorporate scientific perspectives, it may have less to do with the credibility or legitimacy of the supplied information and more to do with scientists entering into repeated personal interactions and conversations with decision-makers. By adhering to the assumption that information, rather than relationships, leads to 'better' decisions, knowledge transfer efforts become unnecessarily convoluted. Consequently, excessive time and resources are often spent on improving information when they would often be better spent on improving relationships.

However, as noted in a review by Beratan (2007), many research organisations are reluctant to invest in relationship-building activities because they are resource intensive and risky. And because the outcomes of such activities are difficult to predict or measure—not least because we often pick the wrong metric—such investments are difficult to justify in today's output-focused organisational culture. Even though most agencies know impersonal communication has limited impact, they still tend to direct their investments into simple low-cost 'products', such as fact sheets, websites and newsletters, which can be pointed to and measured as concrete outputs (Beratan 2007).

The costs of this error in logical typing do not, however, stop at misdirection of resources that could have greater impact elsewhere. Indeed, this is perhaps the least of its ill effects. As we have seen, errors in logical typing create paradox. This leads to errors in problem formulation, which in turn inadvertently generates solutions that paradoxically reinforce or escalate the problem at hand. As per Bateson (2000) and Watzlawick *et al.* (2011), this is the path by which seemingly intractable 'wicked' problems are born. In the following sections, the perverse paradoxical outcomes generated by errors in logical typing are explored in relation to both knowledge transfer and adaptive capacity.

Knowledge transfer paradox: entering the vicious circle

When scientists say they would like to see their research used by managers and decision-makers to make *better informed* decisions, what scientists really mean is that they want a *say* in those decisions, to influence how reality is unfolding, based on what they know.[12] The mainstream scientific paradigm, however, prohibits the scientist from stating this explicitly, as doing so compromises their privileged stance of objective observer—that of standing outside the system and looking in. Thus, the term 'better informed' is a linguistic device, an abstraction scientists subconsciously use to maintain their guise of objectivity, and to distance themselves from the "arena where human beings struggle with one another" (Ong 2002: 43). This is a consequence of how scientists have been socialised—to enshrine objectivity and to value only representational knowledge that reflects external reality. Thus, in the name of 'objectivity', scientists adhere to representational-referential[13] forms of communication,[14] which passively represent reality for others to understand. Then they hope these passive representations will be picked out from the chaotic milieu of rhetoric, power and persuasion shaping everyday reality—and rationally be given privilege in decision-making, by virtue of the credibility and accuracy of their depictions.

This 'strategy of hope' (as per Roux *et al.* 2006), however, is only valid if the string of assumptions outlined earlier in the chapter hold. As we have seen, there are strong arguments against these assumptions in favour of an alternative complexity-based paradigm. From a complexity perspective, decision-making is better understood as a process of interacting with other systems, rather than individual information processing. Hence, the term 'better-informed decisions' is oxymoronic, and its pursuit a diversion that keeps scientists from achieving their real objective: *influence*. Ironically, science is thus neutered by its own paradigm. As outlined in our preceding discussions, if scientists want their research to have greater impact on how reality unfolds, they best enter the 'struggle' and expand the breadth and scope of their personal interactions to include decision-making domains. This relates to Pielke's (2007) arguments that the work of scientists who are disengaged from the policy process (e.g. the 'pure scientist' or 'science arbiter') is only effective in policy situations where there is already broad consensus and low uncertainty. Thus, the cliché 'you gotta be in it to win it' applies.

As with most things, however, this is more easily said than done. Coherent patterns of behaviour in complex systems resist change, and in human systems this translates to powerful interests working to dominate discourse in order to maintain the status quo, as per Foucault (1980).[15] Thus the 'neutering' of science is self-reinforcing—by way of research programs that relegate scientists to the role of supplying information and tools for 'decision-support', that measure academic performance in terms of information outputs such as reports and journal articles, and which prohibit scientists from expressing opinions or entering into policy debates[16]—and all the while increasing demands that scientists demonstrate greater 'research impact'.

Once again, we have a paradoxical situation created by an error in logical typing—assuming that our clockwork maps of a dynamic, interconnected and emerging territory are more than just navigation tools, that they *are* the territory, despite our experiences to the contrary. This particular paradox also presents a classic double bind[17] situation, whereby by the rules of a system are such that success in meeting one demand results in failure to meet another, and vice versa. In this case, if the scientist maintains a purely objective stance, he or she will likely fail to achieve substantive research impact. However if research impact is achieved, he or she will have failed to maintain their objectivity. In psychology, double binds are notable for creating impasses and provoking anxiety, neurosis and confusion in those affected; indeed they are frequently employed as a means of subtle control over others (Watzlawick *et al.* 2011). According to Watzlawick *et al.* (*ibid.*), a key feature of double binds is that the people caught up in them are usually unable to discern the nature of the paradox they are facing. As a consequence, they fail to realise that their problems are generated by the rules or premises of the system itself. Thus any attempts to resolve the bind via logical analysis *within* the framework of the system (i.e. single loop learning), will only reinforce or escalate this situation, and so a vicious circle is born.

A highly simplified hypothetical summation of the climate change debate can serve to illustrate how a vicious circle is born of an unchallenged underlying assumption in relation to knowledge transfer. As per the string of Newtonian assumptions outlined earlier in the chapter, many scientists assume that presenting a sufficiently persuasive argument based on the rational facts should influence the beliefs, decisions and behaviours of others. From this assumption, scientists concerned with the lack of action around climate change present a barrage of facts to the public and decision-makers. However, the scientists do not realise that information alone rarely affects people's behaviours, or that information that conflicts with people's existing beliefs has the polarising effect of causing some people to change their beliefs, while causing others to reject the new information, discredit the messenger and further entrench their existing positions, as per the backfire effect of cognitive dissonance (Nyhan and Reifler 2010). Because these scientists have not undergone double-loop learning that recognises or challenges their underlying assumption that facts change people's beliefs and behaviours, they fail to see that presenting climate facts created much of the climate denial and scepticism in the first place. In absence of this insight, scientists then respond to mounting climate scepticism by presenting *more* contradictive facts (i.e. they apply single-loop learning). Thus, in absence of double-loop learning, the scientists escalate their error by intensifying a strategy that inadvertently increases rejection of climate science among those they are trying to convince. Therefore, they find themselves entrained in a vicious circle. This phenomenon is likewise documented by Beratan's (2007, citing other therein) review, which states: "In fact, additional scientific facts may reinforce value disputes and competing, and more information may lead to more confusion rather than less".

Adaptive capacity paradox: an error in problem formulation

Complexity thinking reveals another concern with how environmental management problems have been formulated—this time in relation to the textual notion of *adaptive* as it is applied to resource management and institutional capacity. As detailed in Chapter 1, adaptive management involves cyclical learning and policy adjustment and implementation. Hence, adaptive management is environmental management's equivalent to Lewin's action research cycle in organisational development and Drucker's management-by-objective system in organisational management (see Chapter 6). All three involve designating a desired system objective, then applying a plan–act–reflect cycle of negative feedback to stabilise the system around the selected objective, using participatory methods. Attempting to apply this approach to living self-organising systems, however, is an error in logical typing, as it assumes these systems can be designed and controlled cybernetically, so as to maintain an externally determined system state. This confuses self-organising living systems with mechanical self-regulating systems. As outlined exhaustively in Chapter 6, human systems are dynamically self-organising, and their pattern is a consequence of the co-evolution and harmonisation of a vast multitude of internal and external interactions. The primary imperative of any self-organising system is to maintain its identity,[18] its existing pattern of behaviour. Any attempts to stand outside and design or tinker with such systems so as to make them 'more adaptive'—or to meet any other external prescription—are likely to meet with either frustration, as the system absorbs any attempts to alter it, or with dismay, as it reacts in unpredictable and potentially undesirable ways. Curtin (2014) similarly argues that "Adaptive responses in and of themselves are intrinsically halfway solutions that do the 'wrong thing righter' by attempting to plan rather than design for the future". This perhaps explains why failure rates of 70–80 per cent are reported in the literature for both adaptive management projects (Medema *et al.* 2008; Walters 1997), and for change management initiatives in the business world (Burnes 2005; Senge *et al.* 1999).

However, as is the case for knowledge transfer, the cost of this error in logical typing goes beyond just getting poor return on invested resources. In the case of wicked problems, which span multiple system boundaries and domains of governing order, it is the overarching framework of the system—with its accompanying objectives, rules and premises[19]—that is actually responsible for generating the problem situation. Without changing this overarching framework, using adaptive management to tinker with (i.e. cybernetically adjust) the behaviours of any of the parts of the larger system in which the problem is embedded (i.e. doing more, less or variations of the same), will paradoxically reinforce and possibly escalate the problem at hand, as per Galois's group theory. Analytical policy adjustment and cyclical learning amount to single-loop learning (as per Argyris and Schön 1996), and will probably only generate first order change (as per Bartunek and Moch 1987).

On the other hand, if adaptive management targets the rules and objectives of a system's overarching framework (i.e. second-order change, via double-loop learning), efforts will be met with distrust and resistance (Argyris and Schön 1996; Brown and Duguid 2000; Stacey 1996). As noted earlier, this is because second-order change threatens the higher-order governing pattern that constitutes a system's identity. Indeed, concerted attempts at creating second order change may provoke embarrassment and defensiveness among a system's members, and inadvertently generate counterproductive anti-learning patterns (Edmondson and Moingeon 1999) and entrenchment of existing behaviours, as per cognitive backfire (Nyhan and Reifler 2010). As the systems theorist Donella Meadows (1999: 19) observes, the 'higher' the leverage point used to try and change a system, "the more the system will resist changing it—that's why societies tend to rub out truly enlightened beings". Many of the barriers to building adaptive institutions summarised in Chapter 1 are the consequence of homeostatic forces at work in coherent systems. Thus, because adaptive management constitutes an error in logical typing, it may therefore either be relatively ineffective or paradoxically reinforce the problem it is trying to correct, inadvertently creating a vicious circle.

The ultimate aim of environmental management is to see that the earth's biophysical resources are used and managed sustainably. This means they will not be depleted in the long term, thereby ensuring our own long-term survival. As outlined in Chapter 1, it is assumed by many that this can be achieved via *adaptive governance*—improving the *adaptive capacity* of our institutions so as to make them more resilient. By this it is assumed that human systems will be better 'adapted' to their environments if there are strong feedback loops between society and the environment (Folke *et al.* 1998; Hanna *et al.* 1996), allowing for adaptive management (Berkes *et al.* 2003). It is then presumed that detecting and responding to environmental feedback (adaptive management) requires knowledge and understanding of resource and ecosystem dynamics (Folke *et al.* 2005). Hence, it follows that knowledge transfer is a feedback mechanism that can help confer institutions with adaptive capacity. This in turn helps human systems operate in balance with the environment, which then helps us avoid depleting our biophysical environment to our long-term detriment. These were my own assumptions when I found myself puzzling over the disappointing results of the knowledge transfer project that inspired this book.

From a complexity-based ontology, however, there is no outer reality to which an individual system must adapt. To treat the 'environment' as a pre-given background entity is an error in logical typing.[20] What we distinguish as a system's 'environment' is actually the vast fabric of nested dynamic interactions, of which the system is a part, that together are perpetually co-creating reality.[21] As outlined in Chapter 5, any sense of separation between the system and its 'environment' is an illusion created by the scale of observation. The main imperative of these networked systems is maintaining their circular pattern of organisation (i.e. identity) while they interact—they have no interest in sustaining their 'environments' as no such thing exists. Reality is the pattern of jostling tension that emerges from their conflicting and cooperating identity-conserving interactions.

The greater the number and diversity of couplings between these systems, the larger the domain of possible interactions they may enter into (Chapter 6). This in turn increases the domain possible futures they may create. This is better understood as conferring the system with *plasticity* or *creative capacity* (via two-way interaction), rather than conferring it with adaptive capacity—which denotes its capacity to *react or change* (via one-way interaction).

Plasticity and decoupling

As described in Chapter 6, being able to use language to share and coordinate our imaginings of the future has enabled humans to enter into increasingly complex and ordered patterns of interaction. The result is an increasingly complex society, wrought by our increasing connectivity. This has created vast and complicated global supply chains that have 'strung out' the feedback loops that connect members of industrialised societies with life-sustaining biophysical resources. We are no longer pre-occupied with basic survival. This has allowed our every-day attention to largely shift away from the biophysical environment—most of us no longer directly couple or interact with many aspects of it. Instead, our attention and interactions—aided by technology and the Internet—are now focused on increasingly abstract social imaginings and constructs, which become self-sustaining over time. Because attention[22] is like fertiliser—it makes phenomena grow—this serves to further increase the domain and complexity of higher order socio-economic governing patterns. Which in turn further separates us, spatially and temporally, from the biophysical resources that sustain us.

These higher order socio-economic patterns are so plastic and resilient—by virtue of their complexity and global connectivity—that they have, so far, spontaneously resolved strains on resource supply. They do so by creatively intensifying or shifting resource extraction via technological and social innovations. This has allowed members of industrialised economies to evade the constraining negative feedback that would succeed in entraining/harmonising more simply-ordered social systems into the circular governing patterns of biophysical systems.

Adaptive approaches to management and institution building aim to artificially bring about this entrainment/harmonisation by re-establishing more direct cybernetic feedback loops between human society and the biophysical environment. This is often conceived as improving the supply of scientific information for decision-making, or designing institutions to incorporate scientific knowledge in cyclical learning and decision-making cycles. However, as discussed above, artificially engineered information-based feedback loops are not likely to succeed in short-circuiting the complex higher order governing patterns that drive societal behaviour, and may indeed exacerbate the problems they are intended to resolve. Human thinking and behaviour is not driven by conscious information processing and rational decision-making, as is assumed by the Newtonian mind-as-a-computer paradigm. As detailed in Chapter 6, behaviour, communication, language, learning, imagination and all other social phenomena are just extensions of dynamic embodied cognition—stable higher order patterns of interaction that spontaneously emerge from iterative couplings between living systems (Maturana and Varela 1992).

Once again, it is relationship not information that drives system behaviour. When we attribute the erosion of sustainability to poorly-informed decisions, or poorly designed institutions, we commit logical errors in problem formulation. Global sustainability issues are part of the dynamic pattern emerging from a vast complex of historic interactions, of which we are part. Assuming that the behaviour of large, complex self-organising systems can be deliberately changed by design, prescription or information is an error in logical typing. Attempting to do so spawns paradox-generating strategies, such as knowledge transfer and adaptive management.

The escape route: reformulating the problem

> The way out is through the door. Why is it that no one will use this exit?
>
> (Confucius)

A typical, or 'normal', response to knowledge transfer and adaptive management failures would be to assume that the theoretical framework behind these notions is correct, and to rationalise the disappointing results as the consequence of insufficient effort, poor project design, and/or 'contaminating' effects from outside forces. Future energy would be geared toward adhering even more closely to the theoretical framework (i.e. doing more or less of the same)—such as intensifying the output of information to decision-makers, and improving its credibility and usefulness. But this takes us back to Watzlawick *et al.*'s (2011) observation that "failure to resolve a problem doesn't lie in the impossibility of the task, but in the attempted solution" (*ibid.*: 25). The hypothetical examples outlined above show how 'textual' notions of adaptive capacity and knowledge transfer cause logical errors in problem formulation, with paradoxical results. We can now see that the common failures encountered when trying to increase the application of science by decision-makers are quite possibly a result of applying knowledge transfer as the attempted solution.

Thus, the high failure rates encountered in such initiatives should prompt us to seriously consider Watzlawick *et al.*'s observation and 'dig deeper' instead, to explore the unacknowledged assumptions and premises underlying the problem framework. This approach is no doubt the path less travelled. And given its convolutions, uncertainties, risks and challenges, it is understandable why 'travellers' in the scientific disciplines seldom tread it. However, as the previous discussions in this chapter demonstrate, neglecting that path "bent in the undergrowth" (Frost 1916) has potentially serious consequences, the generation of intractable or wicked problems among them. It also creates methodological concerns.[23] For example, when experimental results fail to support a particular theory, often times contaminating effects—the unforeseen and therefore uncontrolled variables believed to disrupt the workings of the theorised system's behaviour—rather than the theory itself, are blamed. A frequent response is to further isolate the system being tested, to close it off from these 'contaminating' variables.

But this can be a perilous approach when dealing with complex systems, as it fails to acknowledge that the behaviour any given complex phenomena is the *emergent* result of a multitude of interacting variables, which cannot be abstracted from surrounding systems without disrupting the behaviour itself (see Funtowicz and Ravetz 1993). Trying to explain unexpected behaviours in complex systems by further isolating them creates a *reductive spiral*, whereby the patterns of relationship that give a system behaviour 'life'[24] are paradoxically severed by attempts to better understand that behaviour by breaking the system down into pieces[25]. This is yet another case of confusing the map with the territory (confusing the theory with the *actual* pattern of interactions it is meant to represent)—and the methodological equivalent of trying to understand the behaviour of an organism by dissecting it, therein causing it to die. Capra (1997: 29) points out the limitations of reductionist methods when dealing with the properties of living systems, properties that are "destroyed when the system is dissected either physically or theoretically, in to isolated elements". This argument is likewise made by Maturana and Varela (1980), who also note that reductive analysis of living systems is limited by the cognitive domain of the observer—the entirety of interactions that comprise a living system are not apparent to the observer, as the observer's cognitive domain, is necessarily different to that of the system he or she is observing.

The theoretical explorations of this book have attempted to separate the old reductionist Western 'map' of reality from the territory it was meant to represent. The reader has been invited to venture across theoretical and philosophical ground unfamiliar to the average mainstream scientist, and in doing so, crumple up their old map of reality and draw new a new one. With this new complexity-based map in hand, how then can scientists avoid paradox and apply their research so as to have greater influence on decision-making?

Shotter (1994, citing others) notes that when we become entrapped by our assumptions, or habits of thought, they are extremely difficult to escape, as they prevent us from recognising proof that our beliefs may be false or inaccurate. Watzlawick *et al.* (2011) observe that people who manage to escape double binds or otherwise paradoxical situations, usually do so through sudden insight. This insight typically appears as a 'jump' to a weird or counter-intuitive solution, which often applies itself to previous attempts at solving to the problem. Indeed, this was the case for me, when a sudden insight around the scale dependence of chaos and its unifying effect helped me frame the complexity-based ontology I synthesised in Chapter 5. As discussed earlier, this 'jump', or insight, *provides a route out of the system*, allowing those caught to escape the framework of rules, premises and assumptions responsible for generating the problem situation to begin with. This notion of 'mentally escaping' a system's boundaries also resonates with Shotter's (1994) argument that such entrapments are not caused by the true nature of ourselves or reality, but rather by the literary and textual ways we have formulated our problems.[26] For example, scientists wanting to escape the double bind presented earlier can do so, either by (i) accepting that it is indeed a "participatory universe" (Wheeler 1973: 244), thereby abandoning their objectivity in favour of connecting with others and adding their rhetoric and vision to the conversation, or by (ii) maintaining their objective stance, and abandoning their attempts to improve their research impact by way of throwing information at problems.

Thus, the problem of bridging the gap between research and management can perhaps be resolved by reformulating the problem, so as to remove the assumption that applying science involves rational information processing on the part of decision-makers (and therefore the requirement that scientific knowledge be successfully transferred *to* them). Hence, rather than asking why knowledge transfer efforts in the environmental field so often fall short of improving the adaptive capacity of institutions, the problem can be reformulated from a complexity-based perspective to ask instead: "how can environmental scientists have greater influence on the emerging pattern of reality, such that human systems are more sustainably coupled with biophysical systems?"[27] By replacing our Newtonian assumptions with those of a complexity-based ontology, we can see that active relating, not information, is what drives behavioural patterns—and that decision-making is a process of coordinated interaction between people, rather than individual information processing.

And so it was that I re-examined the disappointing results of the knowledge transfer project that had spurred me onto this journey of solving the riddle of knowledge transfer (see Chapter 1). A main aim of the project had been to change people's levels of (representational) knowledge about the area, in terms of its ecological sensitivity and sustainable management. However, because I had based my project evaluation framework around Park's (1999) tripartite knowledge typology, I evaluated how the project changed people's patterns of relating, as well as their levels of representational knowledge. The evaluation results showed that despite intensive stakeholder engagement, the knowledge transfer process had failed to have much effect on the representational knowledge of local stakeholders—this was the disappointing part. What *had* changed, however, were people's patterns of relating. The evaluation showed that the stakeholder engagement components of the knowledge transfer effort had catalysed positive relationships between stakeholder groups in the region, and between regional stakeholders and scientists. Through this improved and expanded network of relations, these groups had begun conferring and cooperating on a variety of resource management issues and projects. Chapter 6 outlines how the 'knowledge' of a system is better understood as a function of its internal and external connectivity (i.e. relationships between its agents, which confer the system with *creative capacity*), than as the amount of information held inside the heads of its individual members. Thus, when reframed from a complexity perspective, the knowledge transfer project no long constitutes a failure. Believing it so was a classic example of an error in problem formulation. Rather, the increased connectivity between stakeholder groups and scientists can be viewed as having succeeded in enhancing the *creative capacity* (as opposed to adaptive capacity) of study area's socio-ecological system.

In the following sections I tentatively propose some complexity-based strategies for addressing the reformulated problem of how environmental scientists can influence the emerging pattern of reality such that human systems are more sustainable. I do so cautiously, and with some misgiving, as I believe there can be no prescriptive solutions when dealing with complex systems and wicked problems. As Richardson (2008: 22) states, "Complexity 'thinking' is a particular attitude towards our ideas about the world and the world itself, not a particular tool/method, or even a particular language". From his perspective, managers will gain more from understanding the *philosophy* of complexity, and the light it sheds on the limits of knowledge and on how underlying assumptions shape our worldviews and decisions (*ibid.*). This resonates with Brockman's insightful comments about the extraordinary and ground-breaking work of Gregory Bateson:

> Bateson is not clearly understood because his work is not an explanation, but a commission. As Wittgenstein noted, "a commission tells us what we must

do". In Bateson's case, what we must do is reprogram ourselves, train our intelligence and imagination to work according to radical configurations.

(Brockman 1977: 6)

Imagination: coupling with the future

As discussed earlier, the increasing complexity of global supply chains has 'strung out' the feedback links between industrialised societies and the biophysical systems that sustain them. It appears unlikely that engineered, information-based feedback loops, such as those applied in adaptive management, will be effective at shortening these attenuated linkages. Spatial shortening of these feedback loops will likely only occur 'naturally', in the event of a societal collapse that disorders global socio-economic governance patterns. In light of these difficulties, the other option is to try and establish longer-range temporal coupling with a *desired future state.*

As detailed in Chapter 6, the future *is* imagination. In sharing our imaginations of the future, and therefore how we act into it, we co-create reality. On this basis, in order to influence emerging reality in a significant way, scientists have to participate in it, connect with others, and add their visions of a desired future (e.g. one with well-managed biophysical systems) to the collective's imaginings. From the interactions that follow, new governing patterns might then emerge that somehow reflect their scientific visions. This pertains to the words of Maturana and Varela (1992: 245), who argue that the world we see is "not *the* world but *a* world we bring forth with others". Thus the "world will only be different if we live differently" (*ibid.*). Likewise, Stacey *et al.* (2000: 177) note that "people together shape what they perceive and hence the context in which they jointly act. They do so by pointing to this and inviting people to look at that".

Creative capacity versus adaptive capacity

With this in mind, how then can scientists best share their imaginations so as to influence the emerging pattern of reality, to help build the future? As described in Chapter 6, we do not so much gather and process information about a pre-given reality, then make conscious decisions to better re-arrange it in our favour (i.e. to improve our adaptive capacity). Rather, we *specify* reality via coupling with other systems. Life constructs its own future. For humans, as for all other living organisms, the future is radically unpredictable. Through our own interactions we are co-creators of reality, neither engineers nor designers of it. As Rose (2003) states, we have the ability to construct our own futures, but not under circumstances which we choose. All we can do is add our voices and visions to the milieu, and hope to have influence. As Stacey (2001) argues, instead of trying to find ways of designing and managing whole systems, which no one individual or group can realistically control, the emphasis should instead be on paying attention to *one's own local* participation.

Thus, instead of devising rational objective arguments for others to use in rearranging or designing human systems (so as to improve the 'adaptive capacity' of these systems in relation to some external environment), environmental scientists should perhaps shift their attention towards improving their own *creative capacity*. This means sharing their imaginings via improving the impact of their local participation,[28] and therefore helping *create* a reality that it is more sustainable. This requires that environmental scientists reframe their role of objective observer into that of *active participator*.[29] The following paragraphs outline some possible suggestions for improving the *creative capacity* of scientists and research programmes.

Increase number, diversity and intensity of personal interactions

Creative capacity refers the domain of possible futures a system can create. This is a function of the number and diversity of its connections between its agents and other systems, and pertains to the quality and dynamics of human interaction in the present, as per Stacey's (2001) theory of complex responsive processes. Thus, environmental scientists can increase their creative capacity by expanding their domain of personal interactions. In this way, they are able to add their 'biophysical rich' imaginings of the future to that of others, and through dialogic interaction transform each other and the institutions they comprise (e.g. Healy 1997). New couplings between scientists and areas of the decision-making system can potentially interrupt habitual patterns of thinking and interacting. Scientists may want to be strategic in terms of assessing what scales and domains of governance they should be targeting for their interactions (as per Bodin and Prell 2011), based on the influence they are hoping to achieve. Direct personal coupling, resulting in sustained interaction, is more powerful than indirect means, such as by supplying people with information in the hope that it will rationally convince them to change their existing behaviour pattern. Scientists can also be strategic in their interactions by paying attention to patterns of power relations and ideologies, who is talking/included and who is being silenced/excluded, how people manage their anxiety, and whether and how conversations are spontaneous and excited, or dull and repetitive (as per Stacey 2001). Research programs looking to have impact in a particular area can also use network analysis to identify 'functional holes' in their relationships with stakeholder groups and decision-makers, then strategically focus their efforts on bridging these gaps (see Sandstrom 2011; Syme *et al.* 2012).

We must also recognise that scientific knowledge cannot be effectively disembodied from scientists and applied by others without significant diminishment of meaning, context and influence on decision-making and behaviour. Thinking otherwise is a relic of Taylor's scientific management, and reliant on defunct information-processing views of cognition and decision-making. If science is to have greater influence, it has to be part of the decision-making system in terms of actively relating *bodies*, not just reports. This premise is reflected in Bodin and Prell's (2011) relational approach to resource governance, with its emphasis on social networks; and to Tompkins and Adger's (2004) and Newman and Dale's

(2005) arguments that social networks build resilience and adaptive capacity in socio-ecological systems. It is also reflected in Funtowicz and Ravetz's (1990, 1991, 1993) conception of publicly engaged post-normal science.

Engage in everyday conversation

Conversation is characterised by a creative and uninterrupted flow of cyclical stimulus-response between the parties involved, one that has an immediate effect on their patterns of interaction. Ordinary free-flowing conversation is how we share our imaginations and shape reality through our day-to-day interactions. People are more responsive to conversation than they are to the written word, and they are also more likely to respond to rhetoric-responsive conversation, the language of persuasion and imagination, than they are to passive and 'rational' representations of reality (see Donaldson 2005; Shotter 1994). By relegating their imaginings to passive paper abstractions, divorced from the creative power of day-to-day conversation, rhetoric and personal interaction, scientists limit themselves to a strategy of hope (as per Roux *et al.* 2006)—hoping that someone will seek out their paper imaginings, be persuaded by their rational arguments, and reengineer the system accordingly. Every day conversation has more creative power than the representational-referential written communication typically applied by scientists in their efforts to be objective. This is the premise of Stacey's (2001) complex responsive processes approach to organising. Scientists wishing to have more influence can try to create space for having informal conversations with a diversity of people (as per Shaw 2002), particularly those within decision-making spheres they wish to influence. This may require a restructuring of job descriptions and resources within research institutions.

Seek opportunity in crisis

When adding their conversations to the milieu, scientists can also opportunistically and strategically look to interact with highly connected people who are operating in the areas of greatest connectivity and instability, i.e. where the 'action' is. Unstable 'zones' near the edge of chaos, or near crisis, are where new voices are least likely to subsumed by well-established governing patterns that are highly change resistant (as per Stacey 1996). Thus, these parts of a system are where ideas have the best odds of being amplified and where transformational change is most likely to occur. Instabilities are also created when top managers are replaced in organisations (Nystrom and Starbuck 1984). Such occasions can be turned into opportunities for creating new patterns of interaction between scientists and decision-making bodies. These ideas relate to the resilience literature, which identifies ecological crises (e.g. destructive wildfires, collapse of a fishery) as opportunities to build new and creative socio-ecological relationships (e.g. Gunderson 2003). It should be cautioned, however, that such strategies require tolerance for risk and uncertainty (as per Walters 1986), as the outcome of interactions with others, particularly in zones of instability, cannot be predicted and may not be as desired.

Use narrative to convey a compelling vision of the possible

A number of authors have argued for using rich narrative[30] approaches to understanding and resolving problems in complex socio-economic systems (e.g. Robertson *et al.* 2000; Tsoukas and Hatch 2001; Waltner-Toews *et al.* 2005). A narrative is an "account of events occurring over time" (Bruner 1991: 6). According to Bruner (2009), narrative is used to give experience meaning—it is grounded in personal experience, contextual, and made effective by way of good story-telling. As such, it typically used to create a gripping account of *past events* that conveys a particular worldview of 'what is' and why it is so (see also Leach *et al.* 2010). For example, Meppem and Bourke (1999) illustrate how the sustainability debate is shaped by contesting narratives[31] jostling for rhetorical dominance, each one disputing the others' 'truth claims', which Bourke and Meppem (2000) describe as valorised expressions of conflicting values.

In Chapter 6, we saw that capturing people's imaginations requires a vision for the *future*. This involves persuasion by telling captivating stories about a possible future that excites the imagination, rather than by way of argument through analysis or representation.[32] Using narrative modes of discourse to paint a compelling vision of the *future* as well as to rationalise the past (surfacing and sharing multiple narratives of the past being an important step in revealing underlying motivations and assumptions of different stakeholder groups,[33] and building mutual understanding and trust between them)[34] may help scientists and/or research programmes shift people's attention away from intractable and polarising debates around *what is true*, towards more creative and perhaps constructive dialogue around *what is possible.* Lessons can also be taken from the field of human ecology, which "seeks to imagine what it might be like to live and do well in a humane, sustainable and worthwhile world and to invite and encourage broader community commitment to work towards realizing those futures" (Dyball 2010: 274).

Participate in collaborative planning

Scientists can also enhance their creative capacity in more formal ways, by participating in collaborative planning processes. Collaborative (or communicative) planning[35] (e.g. Dryzek 1997; Innes *et al.* 2007; Meppem and Bourke 1999; Wondolleck and Yaffee 2000) focuses on facilitating communicative interaction between stakeholders, deliberating about collective futures, and fostering community empowerment (Huxley and Yiftachel 2000). As such, it presents scientists with opportunities to interact and build relationships with a diversity of stakeholders, and to share their visions for the future. Because collaborative planning is interdisciplinary, it allows participants—including scientists—to collectively explore a much larger phase space than would be available to them on own, thereby radically expanding the domain and complexity of interactions they can enter into and the possible futures they can co-create. This notion is likewise reflected in the resilience literature, where connecting individuals and organisations across multiple network levels is seen as essential for adaptive governance (e.g. Folke *et al.* 2005; Olsson *et al.* 2007) and the adaptive co-management of resources (e.g. Armitage *et al.* 2009; Plummer *et al.* 2012).

Because collaborative planning increases connectivity[36] (in terms of number and diversity of connections) between actors in a socio-ecological system, it also has transformative potential—the potential to spontaneously generate new patterns of interaction between its participants (as per Stacey 1996). In turn, opportunities are presented for transcending wicked problems caused by polarised views and entrenched patterns of interaction between system agents. This relates to Meppem and Bourke's (1999) "idea that a worldview constructed and reinforced within the bounds of a self-referential discipline limits the potential for creative stimulation and consequent effective action". Meppem and Gill (1998: 121) likewise state: "The conventional stranglehold of disciplinary thinking in policy circles limits the capacity of our decision makers to unravel the complexity of all real world environmental policy and management problems".

Research programmes seeking to improve their research impact and contribute to resolution of complex resource management issues can also benefit from integrating their activities—ideally from design to implementation—with collaborative planning processes. However, as noted by Schusler *et al.* (2003), without the commitment of institutional, financial and human resources for sustained and on-going collaboration (5–10+ years according to Roux *et al.* 2006), deliberative processes are unlikely to result in action on the ground. Unfortunately, because collaboration is very resource and time intensive few such projects exist in real life, and those which do are often unsuccessful in terms of clearly improving resource management (e.g. Curtin 2014; Plummer *et al.* 2012). Chapman *et al.* (submitted) outline a framework for determining appropriate levels of stakeholder engagement when designing and implementing knowledge transfer initiatives within the context of more conventionally structured research programs, which aren't necessarily integrated with long-term planning/management projects, due to practical constraints. Roux *et al.* (2006) suggest that research programs can improve their impact by supporting *communities of practice*:[37] informal and self-organising "groups of people who share a passion for something that they know how to do and who interact regularly in order to learn how to do it better" (Wenger 2004: 2).

Foster creative dialogue

As we saw in Chapter 6, one cannot understand a system whole through analytic processes (i.e. looking at the parts) alone, and attempting to do so will likely reinforce the existing pattern of the system, despite intentions to the contrary. We can never be certain whether our analytically applied scientific knowledge will help resolve our sustainability issues or create more of what we don't want. Creative, non-reductionist processes are more likely trigger insight into higher order patterns that are maintaining a problem situation, and present novel, *qualitatively* different solutions for transforming these patterns (i.e. transforming the system attractor). As the complexity philosopher Paul Cilliers stated, creative and aesthetic activities aren't just diversions, they open up possibilities for "imagining better, more sustainable futures" (Cilliers 2011: 152).

Insight is a right-hemisphere process that cannot occur when the left-brain is engaged in analysis. It is, however, often *preceded* by a period of sustained and frustrated problem analysis. Analysis involves discussion, whereas creativity involves dialogue, as per Table 6.2. Hence, when our analytic and/or defensive thinking and conversation is suspended and shifts to a more creative and non-judgemental mode instead, the potential for transformational change in the system increases. Informal and engaging group processes are more creative and more likely to engage right-brain processes than formal analytical discussions. Spontaneous humour is another activator of right-hemisphere thinking, and can trigger insight into hidden knots of contradiction underlying problem situations (e.g. Hatch and Erhlich 1993). As such, when interacting and collaborating with others, scientists and research programs may want to consider dedicating time and resources to fostering creative dialogue (including satirical approaches), rather than just to discussion. For example, Meppem and Bourke (1999) advocate communicative approaches to developing environmental policy via reflective processes in *discursive communities.* This relates to Funtowicz and Ravetz's (1993) notion of scientific argument as interactive dialogue rather than formalised deduction. Tackling such approaches may require coaching on suspending judgement, turning discussion into dialogue, and mediating conflict.

Through creative dialogue we can also focus our attention on our imaginations and visions for the future, rather than on devising paper plans for a future based on an analysis of what went wrong in the past—with all the defensiveness and pattern reinforcement that involves. From imagination, action follows. Analytically derived plans are abstractions cut off from the life-giving patterns of relationship they represent. As such they are 'dead things' that are unlikely, on their own, to generate substantive action. However, if one does manage to breathe life into them, they may unexpectedly turn into 'monsters' and reap perverse results (as per Stacey *et al.* 2000), as outlined in Chapter 6.

Use modelling to catalyse dialogue and extend imagination

D'Aquino *et al.* (2003) used intensive role-playing to engage stakeholders in designing computer simulations of resource management problems. The role-playing process generated iterative dialogue and interaction among stakeholders, which improved their collective decision-making capacity and created the necessary conditions for emergence of new resource management practices in real life. Thus, depending on how they are wielded, computer models (e.g. Fulton *et al.* 2011; Jones *et al.* 2011) and role-playing games can potentially be used as extensions of people's imaginations and catalysts for creative dialogue. This requires that modelling projects be designed and facilitated with the explicit intention of fostering ongoing creative interaction between participants, rather than just generating 'accurate' depictions of reality. However, as is the case with other forms of deliberation (Schusler *et al.* 2003), without the commitment of sufficient time and resources for on-going collaboration, participatory modelling projects may struggle to sustain joint action (Fulton *et al.* 2013).

Enlist knowledge brokers or boundary organisations

Realistically, many scientists will lack the time, aptitude, skills and/or willingness to deploy the strategies outlined above. Knowledge brokers possessing the appropriate skills and attributes (see Dobbins *et al.* 2009; Lomas 2007) can be employed to facilitate these strategies within the framework of a research program and/or collaborative planning process, by helping scientists communicate better with non-scientists, catalysing and mediating relationships between different individuals and groups at appropriate scales of governance, and identifying opportunities for applying science, as per the suite of knowledge brokering strategies[38] outlined by Michaels (2009). As such, knowledge brokers reduce the interaction burden that stakeholder engagement can impose on both scientists and decision-makers (Chapman *et al.* submitted).

Alternatively, research programs can adjust their hiring and promotion criteria to select for science leaders with good knowledge brokering skills, and build knowledge brokering into their job descriptions and budgets accordingly. In so doing, these leaders can represent the interests and work of other scientists who prefer to be 'left at the workbench', and also take on the responsibility (and risk) of interacting with higher-power stakeholders or in high-stake political arenas. Boundary organisations (Cash and Moser 2000), or bridging organisations (Hahn *et al.* 2006), can serve a similar function to knowledge brokers on a larger scale. However, the one-on-one personal aspect of knowledge brokering makes it particularly suitable for the purposes outlined above.

Change metrics and incentives

Improving the creative capacity of scientists also requires moving away from incentives and evaluations that primarily measure outputs (reports, publications, etc.) or changes in levels of representative knowledge among target end-users of research. Instead, new metrics need to be developed which reflect the scope and *quality* of interactions scientists/research programs have with other parts of the socio-ecological system (particularly decision-makers and decision-making processes), and any qualitatively new patterns of interaction/behaviour that emerge through their relating. Chapman *et al.* (submitted), for instance, posit Park's tri-partite knowledge typology as a basis for evaluating the effectiveness of knowledge transfer. This typology allows practical appraisal of how people's knowledge levels (in the conventional sense) have changed, how their patterns of relating have changed, and how their assumptions, attitudes and actions have changed. Meppem and Gill (1998) suggest key measures of success for sustainability planning should focus on *context-related processes,* such as the maintenance of creative learning frameworks. They also argue that stakeholders should be part of the science evaluation process, as is likewise contended by Funtowicz and Ravetz (1990, 1991, 1993) in their notion of extended peer communities.

Conduct qualitative research

And finally, more research is needed to test the complexity-based paradigm put forward in this book. Specifically, the above suggestions for increasing 'creative capacity' should be assessed within the proposed complexity framework, to document their effect on emerging patterns of behaviour in socio-ecological systems, and to assess how these outcomes compare to that of conventional knowledge transfer efforts. This requires that scientists step outside the realm of quantitative research, and embrace qualitative approaches that serve to *deepen* understanding of social phenomena (see Morgan and Smircich 1980). Such approaches include action research (especially that which is *participatory*), ethnomethodology, phenomenology, and grounded theory. Scientists and research programs can integrate qualitative evaluations into their work by seeking transdisciplinary research partnerships with social scientists and theorists. Brown *et al.* (2010) outline a theoretical framework and describe practical approaches for undertaking transdisciplinary inquiry.

Conclusion

> If you want to build a ship, don't herd people together to collect wood and don't assign them tasks and work, but rather teach them to long for the endless immensity of the sea.
>
> (Antoine de Saint-Exupéry[39])

When I started the research that inspired this book, my aim was to shed light on why knowledge transfer efforts in environmental management so often fall short. In keeping with Watzlawick *et al.*'s (2011) principles of problem formulation, instead of entering into a detailed problem analysis I embarked on a deep investigation of the assumptions underlying the problem statement itself. Through this investigation the workings of our old Newtonian paradigm surfaced, and I was able to synthesise an alternative ontology based on twentieth-century insights derived from the quantum and complexity sciences and process philosophy. In applying this ontology, the errors in logical typing wrought by our Newtonian beliefs and assumptions were revealed, together with their paradoxical consequences. In particular, it showed how an error in problem formulation may have led to a vicious circle whereby our efforts to solve the problem of knowledge transfer are creating the problem we are using knowledge transfer to solve. Shotter (1994) argues that such entrapments are not caused by the true nature of reality, but rather by the language we have used to formulate our problems, so as to uphold our habitual ways of thinking. In so being, we are not absolutely resigned to the perpetuity of our wicked problems, as they are cages of our own thinking, supposed realities we have talked ourselves into through our conversations.

Instead, we can study how we became entrapped in the first place, as did I in this investigation. Or more importantly, we can take our focus away from producing 'things', or redesigning our human systems to improve their 'adaptive capacity', and shift it toward improving our own *creative capacity* instead—by building relationships, changing our conversations, and expanding our day-to-day personal interactions with others. And in both our dealings with others and in our imaginations, we can momentarily suspend our judgement and let go of our strongest beliefs. Then, in the words of Donella Meadows, by "profoundly, madly, letting go" (1999: 19) we open ourselves to an unmediated experience of the present moment, and the insight and creative imaginings that affords. And in turning our gaze outward, we leave the rarefied world of abstraction, and actively join the rabble in the messy and unpredictable process of co-creating reality, with all the tension, conflict, cooperation and uncertainty that entails.

Notes

1 Francis Bacon recognised that the mind is a 'crooked mirror', and the importance of purging one's mind of its 'idols' (rubbing out its pre-existing assumptions and interpretations) prior to attempting to acquire new knowledge (Klein 2012).
2 Noting however that *probabilistic* prediction then applies, and that chaos and therefore predictability are dependent on the scale at which a system is observed.
3 While recognising that qualitative methods also have limitations.
4 Description increases as a function of axioms, but at a diminishing rate of increase (James 2002).
5 In contrast to *post-normal* science, as outlined by Funtowicz and Ravetz (1993), which applies dialogue with 'extended peer communities', rather than just traditional problem-solving, as a means of quality assuring science when uncertainty and/or decision are stakes are high.
6 By the eighteenth century, philosophers such as Hume and Smith were beginning to discuss ideas about laws governing the complex interactions involved in producing, distributing and consuming goods (i.e. the economy; Hausman 2008). In 1776, Adam Smith published his classic explanation of how individuals in pursuit of their own interests, create unintended consequences (i.e. the 'invisible hand' unintentionally promoting the interests of society; *ibid.*). Thus economics has since been primarily interested in the cumulative consequences of individual pursuit of wealth, and one of its primary branches of inquiry concerns rational choice (*ibid.*).
7 Indeed, the utility model of rational choice is not falsifiable and can be made to fit any given set of events, and is therefore highly limited in its applicability (Hodgson 2012).
8 Refer to Chapter 1 for details.
9 For examples see Pahl-Wostl *et al.* (2007) and Tàbara and Pahl-Wostl (2007).
10 For examples see Dryzek (1997), Meppem and Bourke (1999) and Wondolleck and Yaffee (2000).
11 For examples see Berkes and Folke (2002), Folke *et al.* (2005) and Olsson *et al.* (2007).
12 Which scientists consider to be 'evidence', which leads them to also express a desire for 'evidence-based' decision-making. 'Evidence' is generally understood as *that*

which reveals 'truth'. This understanding clearly relies on Newtonian assumptions of a knowable, external and objective reality. It also reflects the role of evidence in science from the scientific realism perspective. Kelly (2008), however, states that the concept of evidence is inseparable from 'justification'—evidence is that which *justifies belief*, rather than that which *reveals truth*. This understanding is reflected Kuhn's notion of *normal* science, whereby evidence is used to justify paradigms rather than uncover reality. Refer to Kelly (2008) for a discussion on the disputed meaning of 'evidence' as a central concept in the epistemology and philosophy of science.

13 As per Shotter (1994).

14 The impersonalisation and objectification of science has been further amplified by the decontextualisation of concepts afforded by reliance on the written word, as per Donaldson (2005).

15 Foucault (1980) links discourse with power and ideology, arguing that power works through discourse to shape beliefs and attitudes. As such, expert discourse becomes a means of restricting alternative discourse and ways of thinking. Powerful, vested interests attempt to dominate discourse by controlling knowledge through technologies that enable surveillance, enforcement and discipline (*ibid.*).

16 This brings the disturbing trend of scientists being muzzled by democratically elected governments into sharp relief. The recent muzzling of Canadian public scientists is a particularly alarming example (see www.cbc.ca/news/canada/story/2012/03/23/f-federal-scientists.html).

17 As per Bateson (2000).

18 From a complexity perspective a living system's imperative is maintaining its identity—its existing pattern of circular organisation. This is a subtle but important distinction from the Darwinian assumption that a living system's imperative is survival. Thus, we cannot make external determinations that an organisation must change its pattern in order to survive, or fit better with its environment. Because from the organisation's perspective, change is death—the death of its present identity or pattern of organisation, which it will intrinsically resist.

19 Which are socially generated, thereby making wicked problems social problems requiring social solutions.

20 This resonates with Margulis's (1989) arguments against life adapting to a passive environment and Lovelock and Margulis's (1974) Gaia theory.

21 Gunderson and Holling's (2002) concept of *panarchy* captures the complexity-based understanding of systems as dynamic nested interactions, both creative and destructive, but retains a 'flattened' Newtonian focus on system change and adaptation (i.e. systems adapting to reality, rather than co-creating reality).

22 As discussed in Chapter 6, when we become aware of something, we automatically couple with it, and through coupling reality is created.

23 Also see Cilliers's (2005) notion of incompressibility in complex systems.

24 Through emergence, as per Chapter 5.

25 Earlier in this chapter I make the argument that this application of reductionism is wrongfully premised on the Newtonian assumption that cause and effect can be quantified across chaotic system boundaries.

26 This brings us back to Gödel's theorems. Mathematics involves structuring fields of knowledge into deductive systems, which are *intellectual constructs* fully defined by the propositions put forth as the axioms of the field (Askanas 2006). As such, the axiomatic systems referred to in Gödel's theorems are *intellectual constructs* that hamper

full understanding of the object under study (*ibid.*). Thus, the explanatory limits of an axiomatic system are an artefact of how we've *constructed* that axiomatic system, rather than being 'real' or 'true' limits. In other words, a set of axioms are incapable of formally proving some things which are evidently true about the system they describe; this paradox can only be resolved by resorting to non-formal *metalanguage* (Witzany 2011).

27 It should be noted that the structure of this question raises ethical concerns, as it is raised solely from the perspective of scientists' values and interests versus those of society or an affected community. The Newtonian worldview held by most scientists is one whereby reductionist science is seen as objective and value-free, and scientific knowledge is seen as instrumental to increasing societal well-being. Scientific information is not, however, objective and value neutral (Foucault 1980), and scientists should be wary of privileging their values and preferences as surrogates for those of society as a whole (Lackey 2001). In order to avoid privileging certain systems of power over others, scientists need to acknowledge and accept influence from diverse other 'ways of knowing' (Meppem and Bourke 1999).

28 Pielke (2007) suggests that scientists can become actively engaged in policy debates by taking on the roles of 'issue advocate' (taking a definite stand on an issue and preferred policies) or 'honest broker' (providing a wide range of policy options without advocating for any one of them), neither of which focus on sharing imagined futures. Pielke argues, however, that advocacy by scientists politicises and damages the credibility of science. This stance is disputed by some critics (e.g. Howe 2007) who suggest scientific advocates are needed to balance the views of powerful interests and prejudices in policy debates.

29 For years, scientists have battled the Rosenthal Effect—the impact that the opinions, outlooks, theoretical and practical biases of researchers have been shown to have on performance of their subjects, be they rats or humans (Watzlawick *et al.* 2011). Scientists typically view observer bias as negative, and are always looking for ways of eliminating it. However, as quantum and chaos theory have shown us, this is a utopian ideal that can never be achieved. If environmental scientists reframed their understanding of observer bias from that of a problem to that of a potential advantage, they could potentially capitalise on the subjective influence research has on a system's behaviour.

30 Bruner (2009) argues that there are two complementary and irreducible forms of cognitive functioning or thought: *logico-scientific* and *narrative*. These two forms differ radically in terms of how they are verified and what they convince of. Logico-scientific thought is verified by logic and empirical proof, and aims to convince others of 'truth' about reality by way of a good argument. Narrative, on the other hand, is verified by its 'lifelikeness', and convinces others of verisimilitude (truthlikeness) by way of a good or inspiring story. The narrative mode involves imaginative application, "good stories, gripping drama, believable (though not necessarily 'true') historical accounts. It deals with human or human-like intention and the vicissitudes and consequences that mark their course" (*ibid.*: 13). The logico-scientific mode by comparison, is "heartless" in its logic. It seeks to transcend 'the particular' through higher and higher abstraction, and "washes stories away when causes can be substituted for them" (*ibid.*).

31 Bourke and Meppem (2000) argue that averting ecological crisis requires instigating social processes that encourage stakeholders to resist the distortions of political narratives upholding the status quo.

32 This does not discount the need for or importance of scientific analysis; rather it recognises that scientific analysis, on its own, is unlikely to solve complex socio-

ecological problems, and often has limited persuasiveness. Scientists may have more influence on reality-shaping dialogue by sharing (and thereby likely also transforming) their imaginative visions (based on conclusions, values, and aspirations drawn from scientific analysis), than by making analytical arguments.

33 As per Bourke and Meppem (2000).

34 E.g. Chapman (2004).

35 Collaborative/discursive planning focuses on using creative *dialogue* to build trust, relationships and mutual understanding among participants, with the aim of inspiring joint action. More conventional stakeholder planning processes involve negotiation and analytic problem-solving through *discussion*, with participants focused on presenting and defending their positions with the aim of 'winning'. Meppem (2000) advocates the *discursive community* as a genuinely collaborative process, one which is based on learning and the application of transdisciplinarity and methodological pluralism. According to Meppem, the strength of discursive communities lies in their capacity for "establishing relational connections" (*ibid.*: 54).

36 Nudging it closer to the edge of chaos, as per (Kauffman 1993).

37 The community of practice school of thought emphasises the importance of human contact in learning. Rather than being with concerned with the cognitive processes and structures involved in learning, it asks what kinds of social situations allow for learning to take place (Lave and Wenger 2001).

38 Chapman *et al.* (unpublished manuscript) describe the different types of knowledge and outcomes generated by these strategies, using Park's (1999) knowledge typology and Gardner's (2001) framework for determining appropriate levels of stakeholder engagement.

39 A quote in common use and generally attributed to Saint-Exupéry, although the exact source has proved impossible to establish.

References

Argyris, C. and Schön, D. A. (1996). *Organizational Learning II: Theory, Methods and Practice*. Reading, MA: Addison-Wesley.

Armitage, D. R., Plummer, R., Berkes, F., Arthur, R. I., Charles, A. T., Davidson-Hunt, I. J. *et al.* (2009). Adaptive co-management for social-ecological complexity. *Frontiers in Ecology and the Environment 7*(2), 95–102.

Askanas, M. (2006). Gödel's incompleteness theorems: A brief introduction. Retrieved 15 October 2013 from http://math.mind-crafts.com/Gödels_incompleteness_theorems.php.

Bartunek, J. M. and Moch, M. K. (1987). First-order, second-order, and third-order change and organization development interventions: A cognitive approach. *Journal of Applied Behavioral Science 23*, 483–500.

Bateson, G. (2000). *Steps to an Ecology of Mind: Collected Essays in Anthropology, Psychiatry, Evolution, and Epistemology*. Chicago, IL: University of Chicago Press.

Bateson, G. (2002). *Mind and Nature: A Necessary Unity*. Cresskill, NJ: Hampton Press.

Beratan, K. K. (2007). A cognition-based view of decision processes in complex social–ecological systems. *Ecology and Society 12*(1), 27.

Berkes, F., Colding, J., and Folke, C. (2003). Introduction. In F. Berkes, J. Colding and C. Folke (eds), *Navigating Social-ecological Systems: Building Resilience for Complexity and Change* (pp. 1–29). Cambridge: Cambridge University Press.

Berkes, F. and Folke, C. (2002). Ecosystem dynamics and local knowledge. In L. H. Gunderson and C. S. Holling (eds), *Panarchy: Understanding Transformations in Human and Natural Systems*. Washington, DC: Island Press.

Bird, A. (2011). Thomas Kuhn. *Stanford Encyclopedia of Philosophy* (Winter 2011 edition). Retrieved from http://plato.stanford.edu/cgi-bin/encyclopedia/archinfo.cgi?entry=thomas-kuhn.

Bodin, Ö. and Prell, C. (2011). *Social Networks and Natural Resource Management: Uncovering the Social Fabric of Environmental Governance*. Cambridge: Cambridge University Press.

Born, J., Boreux, V. and Lawes, M. J. (2009). Synthesis: Sharing ecological knowledge—the way forward. *Biotropica 41*(5), 586–588.

Bourke, S. and Meppem, T. (2000). Privileged narratives and fictions of consent in environmental discourse. *Local Environment 5*(3), 299–310.

Brockman, J. (ed.). (1977). *About Bateson*. New York: G. P. Dutton.

Brown, J. S. and Duguid, P. (2000). Mysteries of the region: Knowledge dynamics in Silicon Valley. In C.-M. Lee, W. F. Miller, M. G. Hancock and H. S. Rowen (eds), *The Silicon Valley Edge: A Habitat for Innovation and Entrepreneurship* (pp. 16–45). Stanford, CA: Stanford University Press.

Brown, V., Harris, J. and Russell, J. (eds). (2010). *Tackling Wicked Problems Through the Transdisciplinary Imagination*. New York: Earthscan.

Bruner, J. (1991). The narrative construction of reality. *Critical Inquiry 18*(1), 1–21.

Bruner, J. S. (2009). *Actual Minds, Possible Worlds*. Cambridge, MA: Harvard University Press.

Burnes, B. (2005). Complexity theories and organizational change. *International Journal of Management Reviews 7*(2), 73–90.

Capra, F. (1997). *The Web of Life*. London: Flamingo.

Cash, D. W. and Moser, S. C. (2000). Linking global and local scales: Designing dynamic assessment and management processes. *Global Environmental Change 10*(2), 109–120.

Chalmers, A. (2002). *What Is This Thing Called Science?* St Lucia: University of Queensland Press.

Chapman, K. J. (2004). Outcomes of a participatory approach to interpretive planning in the Shark Bay World Heritage Area, Western Australia. Master of Science, Edith Cowan University, Perth, Western Australia.

Chapman, K. J., Horwitz, P., Scherrer, P. and Syme, G. (submitted). Knowledge that acts: A new participatory framework for closing the research-implementation gap. Unpublished manuscript submitted for journal publication in 2015.

Cilliers, P. (2005). Knowing complex systems. In K. Richardson (ed.), *Managing Organizational Complexity: Philosophy, Theory, and Application* (pp. 7–19). Charlotte, NC: Information Age Publishing.

Cilliers, P. (2011). Complexity, Poststructuralism and Organization. In S. Macguire, P. Allen and B. McKelvey (eds), *The Sage Handbook of Complexity and Management* (pp. 142–154). London: Sage.

Curtin, C. G. (2014). Resilience design: Toward a synthesis of cognition, learning, and collaboration for adaptive problem solving in conservation and natural resource stewardship. *Ecology and Society 19*(2), 15.

D'Aquino, P., Le Page, C., Bousquet, F. and Bah, A. (2003). Using self-designed role-playing games and a multi-agent system to empower a local decision-making process for land use management: The SelfCormas Experiment in Senegal. *Journal of Artificial*

Societies and Social Simulation 6(3).

Dobbins, M., Robeson, P., Ciliska, D., Hanna, S., Cameron, R., O'Mara, L., *et al.* (2009). A description of a knowledge broker role implemented as part of a randomized controlled trial evaluating three knowledge translation strategies. *Implementation Science 4*(1), 23.

Donaldson, A. (2005). Writing in organizational life: How a technology simultaneously forms and is formed by human interaction. In R. D. Stacey (ed.), *Experiencing Emergence in Organizations: Local Interaction and the Emergence of Global Patterns* (pp. 142–163). New York: Routledge.

Dryzek, J. S. (1997). *The Politics of the Earth: Environmental Discourses.* Oxford: Oxford University Press.

Dyball, R. (2010). Human Ecology and Open Transdisciplinary Inquiry. In V. Brown, J. Harris and J. Russell (eds), *Tackling Wicked Problems Through the Transdisciplinary Imagination* (pp. 273–284). London: Earthscan.

Edmondson, A. and Moingeon, B. (1999). Learning, trust and organizational change. In M. Easterby-Smith, L. Araujo and J. Burgoyne (eds), *Organizational Learning and the Learning Organization* (pp. 157–175). London: Sage.

Foley, B. and Griffiths, S. (2011). *Engaging behaviour: Behavioural economics and citizen engagement* (Report). London: United Kingdom Department for Communities and Local Government.

Foley, D. K. (2004). Rationality and ideology in economics. *Social Research 71*(2), 329–342.

Folke, C., Colding, J. and Berkes, F. (2003). Synthesis: building resilience and adaptive capacity in social-ecological systems. In F. Berkes, J. Colding and C. Folke (eds), *Navigating Social-Ecological Systems: Building Resilience for Complexity and Change.* Cambridge: Cambridge University Press.

Folke, C., Hahn, T., Olsson, P. and Norberg, J. (2005). Adaptive governance of social-ecological systems. *Annual Review of Environment and Resources 30*, 441–474.

Folke, C., Pritchard, L., Berkes, F., Colding, J. and Svedin, U. (1998). The problem of fit between ecosystems and institutions. Retrieved 17 December 2008 from www.ihdp.uni-bonn.de/html/publications/workingpaper/wp02m.htm.

Foucault, M. (1980). *Power/Knowledge: Selected Interviews and Other Writings, 1972–1977.* New York: Pantheon Books.

Frost, R. (1916). The road not taken. In his *Mountain Interval.* New York: Henry Holt.

Fulton, E. A., Jones, T., Boschetti, F., Sporcic, M., De la Mare, W., Syme, G. J. *et al.* (2011). A multi-model approach to stakeholder engagement in complex environmental problems. Paper presented at the MODSIM 2011: 19th International Congress on Modelling and Simulation, Perth, Western Australia.

Fulton, E. A., Jones, T., Boschetti, F., Chapman, K., Little, R. L., Syme, G. *et al.* (2013). Assessing the impact of stakeholder engagement in management strategy evaluation. *International Journal of Economics and Management Engineering 3*(2), 82–98.

Funtowicz, S. O. and Ravetz, J. R. (1990). *Uncertainty and Quality in Science for Policy* (vol. 15). Dordrecht: Kluwer.

Funtowicz, S. O. and Ravetz, J. R. (1991). A new scientific methodology for global environmental issues. In R. Costanza (ed.), *Ecological Economics: The Science and Management of Sustainability* (pp. 137–152). New York: Columbia University Press.

Funtowicz, S. O. and Ravetz, J. R. (1993). Science for the post-normal age. *Futures* (September), 739–755.

Funtowicz, S. O. and Ravetz, J. R. (2003). Post-normal science. In International Society for Ecological Economics (ed.), *Online Encyclopedia of Ecological Economics*. Retrieved from http://isecoeco.org/pdf/pstnormsc.pdf.

Gardner, S. (2001). Stakeholder relationship management: A strategic tool for communicating and managing sustainable marine futures for WA? Presentation at School of Management, Faculty of Business and Public Management, Edith Cowan University, Western Australia.

Gunderson, L. H. (2003). Adaptive dancing: Interactions between social resilience and ecological crises. In F. Berkes, J. Colding and C. Folke (eds), *Navigating Social-Ecological Systems: Building Resilience for Complexity and Change*. Cambridge: Cambridge University Press.

Gunderson, L. H. and Holling, C. S. (eds). (2002). *Panarchy: Understanding Transformations in Human and Natural Systems*. Washington, DC: Island Press.

Hahn, T., Olsson, P., Folke, C. and Johansson, K. (2006). Trust-building, knowledge generation and organizational innovations: The role of a bridging organization for adaptive comanagement of a wetland landscape around Kristianstad, Sweden. *Human Ecology 34*(4), 573–592.

Hanna, S., Folke, C. and Maler, K. G. (eds). (1996). *Rights to Nature*. Washington, DC: Island Press.

Hatch, M. J. and Erhlich, S. B. (1993). Spontaneous humour as an indicator of paradox and ambiguity in organizations. *Organization Studies 14*(4), 505–526.

Hausman, D. M. (2008). Philosophy of economics. *Stanford Encyclopedia of Philosophy*. Retrieved from http://plato.stanford.edu/entries/economics.

Healy, P. (1997). *Collaborative Planning: Shaping Places in Fragmented Societies*. Vancouver: UBC Press.

Herfeld, C. (2012). Tracing the origins of rational choice theory. Retrieved 4 September 2012 from www.mpiwg-berlin.mpg.de/projects/projects/DeptII_Herfeld_Tracing.

Heylighen, F., Cilliers, P. and Gershenson, C. (2007). Complexity and philosophy. In J. Bogg and R. Geyer (eds), *Complexity, Science and Society* (pp. 117–134). New York: Radcliffe Publishing.

Hodgson, G. M. (2012). On the limits of rational choice theory. *Economic Thought 1*, 94–108.

Holland, J. H. (1987). Complex adaptive systems: A primer. *Santa Fe Institute Bulletin 2*(1), 8–10.

Howe, R. B. (2007). Review of *The Honest Broker: Making Sense of Science in Policy and Politics*, Roger A. Pielke Jr. *Canadian Journal of Political Science 41*(3), 801–803.

Hunt, J. and Shackley, S. (1999). Reconceiving science and policy: academic, fiducial and bureaucratic knowledge. *Minerva 37*(2), 141–164.

Huxley, M. and Yiftachel, O. (2000). New paradigm or old myopia? Unsettling the communicative turn in planning theory. *Journal of Planning Education and Research 19*(4), 333–342.

Innes, J. E., Connick, S. and Booher, D. (2007). Informality as a planning strategy: collaborative water management in the CALFED Bay-Delta Program. *Journal of the American Planning Association 73*(2), 195–210.

James, P. (2002). *International Relations and Scientific Progress*. Columbus, OH: Ohio State University Press.

Jones, T., Glasson, J., Wood, D. and Fulton, E. A. (2011). Regional planning and resilient futures: Destination modelling and tourism development—the case of the Ningaloo coastal region in Western Australia. *Planning Practice and Research 26*(4), 393–415.

Kainer, K. A., DiGiano, M. L., Duchelle, A. E., Wadt, L. H., Bruna, E. and Dain, J. L. (2009). Partnering for greater success: Local stakeholders and research in tropical biology and conservation. *Biotropica 41*(5), 555–562.

Kauffman, S. (1993). *The Origins of Order: Self-Organization and Selection in Evolution*. New York: Oxford University Press.

Kelly, T. (2008). Evidence. *Stanford Encyclopedia of Philosophy* (Fall 2008 edition). Retrieved from http://plato.stanford.edu/cgi-bin/encyclopedia/archinfo.cgi?entry=evidence.

Klein, J. (2012). Francis Bacon. *Stanford Encyclopedia of Philosophy* (Winter 2012 edition). Retrieved from http://plato.stanford.edu/entries/francis-bacon.

Kounios, J., Frymiare, J. L., Bowden, E. M., Fleck, J. I., Subramaniam, K., Parrish, T. B. *et al.* (2006). The prepared mind: Neural activity prior to problem presentation predicts subsequent solution by sudden insight. *Psychological Science 17*(10), 882–890.

Kuhn, T. S. (1962). *The Structure of Scientific Revolutions*. Chicago, IL: University of Chicago Press.

Lackey, R. T. (2001). Values, policy, and ecosystem health. *BioScience 51*(6): 437–443.

Lave, J. and Wenger, E. (2001). Legitimate peripheral participation in communities of practice. In J. Clarke, A. Hanson, R. Harrison and F. Reeve (eds), *Supporting Lifelong Learning: Perspectives on Learning and Teaching* (vol. 1, pp. 111–126). London: Routledge-Falmer/Open University.

Leach, M., Melissa Leach, I. S. A. S., Scoones, I. and Stirling, A. (2010). *Dynamic Sustainabilities: Technology, Environment, Social Justice*. London: Earthscan.

Lemos, M. C. and Morehouse, B. J. (2005). The co-production of science and policy in integrated climate assessments. *Global Environmental Change 15*(1), 57–68.

Lomas, J. (2007). The in-between world of knowledge brokering. *British Medical Journal 334*(7585), 129.

Lovelock, J. E. and Margulis, L. (1974). Atmospheric homeostasis by and for the biosphere: the gaia hypothesis. *Tellus 26*(1–2), 2–10.

Margulis, L. (1989). *Gaia: The Living Earth* (Dialogue with Fritjof Capra, vol. 5). Berkeley, CA: Elmwood Newsletter.

Maturana, H. and Varela, F. J. (1980). *Autopoiesis and Cognition: The Realization of the Living*. Dordrecht: Reidel.

Maturana, H. and Varela, F. J. (1992). *The Tree of Knowledge: the Biological Roots of Human Understanding*. Boston, MA: Shambhala.

McGilchrist, I. (2012). *The Master and His Emissary: The Divided Brain and the Making of the Western World*. New Haven, CT: Yale University Press.

McNie, E. C. (2007). Reconciling the supply of scientific information with user demands: An analysis of the problem and review of the literature. *Environmental Science and Policy 10*(1), 17–38.

Meadows, D. H. (1999). *Leverage Points: Places to Intervene in a System*. Hartland, VT: The Sustainability Institute.

Medema, W., McIntosh, B. S. and Jeffrey, P. J. (2008). From premise to practice: A critical assessment of integrated water resources management and adaptive management approaches in the water sector. *Ecology and Society 13*(2), article 29. Retrieved from www.ecologyandsociety.org/vol13/iss22/art29.

Meppem, T. (2000). The discursive community: evolving institutional structures for planning sustainability. *Ecological Economics 34*(1), 47–61.

Meppem, T. and Bourke, S. (1999). Different ways of knowing: a communicative turn toward sustainability. *Ecological Economics 30*(3), 389–404.

Meppem, T. and Gill, R. (1998). Planning for sustainability as a learning concept. *Ecological Economics 26*(2), 121–137.

Michaels, S. (2009). Matching knowledge brokering strategies to environmental policy problems and settings. *Environmental Science and Policy 12*(7), 994.

Morgan, G. and Smircich, L. (1980). The case for qualitative research. *Academy of Management Review 5*(4), 491–500.

Newman, L. L. and Dale, A. (2005). Network structure, diversity, and proactive resilience building: A response to Tompkins and Adger. *Ecology and Society 10*(1). Retrieved from www.ecologyandsociety.org/vol10/iss1/resp2.

Nyhan, B. and Reifler, J. (2010). When corrections fail: The persistence of political misperceptions. *Political Behavior 32*(2), 303–330.

Nystrom, P. C. and Starbuck, W. H. (1984). To avoid organizational crisis, unlearn. *Organizational Dynamics 12*(4), 53–65.

Olsson, P., Folke, C., Galaz, V., Hahn, T. and Schultz, L. (2007). Enhancing the fit through adaptive co-management: creating and maintaining bridging functions for matching scales in the Kristianstads Vattenrike Biosphere Reserve Sweden. *Ecology and Society 12*(1), 28.

Ong, W. J. (2002). *Orality and Literacy*. New York: Taylor & Francis.

Pahl-Wostl, C., Craps, M., Dewulf, A., Mostert, E., Tabara, D. and Tailieu, T. (2007). Social learning and water resources management. *Ecology and Society 12*(2), article 15. Retrieved from www.ecologyandsociety.org/vol12/iss12/art15.

Park, P. (1999). People, knowledge, and change in participatory research. *Management Learning 30*(2), 141–157.

Pielke, R. A. (2007). *The Honest Broker: Making Sense of Science in Policy and Politics*. Cambridge: Cambridge University Press.

Plummer, R., Crona, B., Armitage, D. R., Olsson, P., Tengö, M. and Yudina, O. (2012). Adaptive comanagement: a systematic review and analysis. *Ecology and Society 17*(3), 11.

Prigogine, I. (1997). *The End of Certainty*. New York: Free Press.

Prigogine, I. and Stengers, I. (1984). *Order Out of Chaos: Man's New Dialogue with Nature*. New York: Bantam Books.

Richardson, K. A. (2005). The hegemony of the physical sciences: An exploration in complexity thinking. *Futures 37*(7), 615–653.

Richardson, K. A. (2008). Managing complex organizations: Complexity thinking and the science and art of management. *Emergence 10*(2), 13–26.

Robertson, M., Nichols, P., Horwitz, P., Bradby, K. and MacKintosh, D. (2000). Environmental narratives and the need for multiple perspectives to restore degraded landscapes in Australia. *Ecosystem Health 6*(2), 119–133.

Rose, S. (2003). *Lifelines: Life Beyond the Gene*. Oxford: Oxford University Press.

Roux, D. J., Rogers, K. H., Biggs, H. C., Ashton, P. J. and Sergeant, A. (2006). Bridging the science–management divide: Moving from unidirectional knowledge transfer to knowledge interfacing and sharing. *Ecology and Society 11*(1), article 4. Retrieved from www.ecologyandsociety.org/vol11/iss1/art4.

Rovelli, C. (2012). Science is not about certainty: A philosophy of physics. *Edge*. Retrieved 15 September 2012 from https://edge.org/conversation/a-philosophy-of-physics.

Sandström, A. (2011). Social networks, joint image building, and adaptability: the case of local fishery management. In Ö. Bodin and C. Prell (eds), *Social Networks and Natural Resource Management: Uncovering the Social Fabric of Environmental Governance* (pp. 288–321). New York: Cambridge University Press.

Schusler, T., Decker, D. and Pfeffer, M. J. (2003). Social learning for collaborative natural resource management. *Society and Natural Resources 16*(4), 309–326.

Senge, P. M., Kleiner, A., Roberts, C., Ross, R. B., Roth, G. and Smith, B. S. (1999). *The Dance of Change: the Challenge of Sustaining Momentum in Learning Organizations*. New York: Currency Doubleday.

Shackleton, C. M., Cundill, G. and Knight, A. T. (2009). Beyond just research: Experiences from Southern Africa in developing social learning partnerships for resource conservation initiatives. *Biotropica 41*(5), 563–570.

Shaw, P. (2002). *Changing Conversations in Organisations: A Complexity Approach to Change*. London: Routledge.

Shotter, J. (1994). *Conversational Realities: Constructing Life through Language*. London: Sage Publications.

Snowden, D. (2003). Complex acts of knowing: Paradox and descriptive self-awareness. *Bulletin of the American Society for Information Science and Technology 29*(4), 23–28.

Stacey, R. (1996). *Complexity and Creativity in Organizations*. San Francisco, CA: Berrett-Koehler.

Stacey, R. (2001). *Complex Responsive Processes in Organizations*. London: Routledge.

Stacey, R., Griffin, D. and Shaw, P. (2000). *Complexity and Management: Fad or Radical Challenge*. New York: Routledge.

Syme, G., Dzidic, P. and Dambacher, J. (2012). Enhancing science in coastal management through understanding its role in the decision making network. *Ocean and Coastal Management 69*, 92–101.

Tàbara, J. D. and Pahl-Wostl, C. (2007). Sustainability learning in natural resource use and management. *Ecology and Society 12*(2), 3.

Tansley, A. G. (1935). The use and abuse of vegetational concepts and terms. *Ecology 16*(3), 284–307.

Todes, D. P. (2009). Global Darwin: Contempt for competition. *Nature 462*(5), 36–37.

Tompkins, E. L. and Adger, N. (2004). Does adaptive management of natural resources enhance resilience to climate change? *Ecology and Society 9*(2), article 10. Retrieved from www.ecologyandsociety.org/vol9/iss2/art10.

Tsoukas, H. and Hatch, M. J. (2001). Complex thinking, complex practice: The case for a narrative approach to organizational complexity. *Human Relations 54*(8), 979–1013.

Walters, C. J. (1986). *Adaptive Management of Renewable Resources*. New York: McGraw-Hill.

Walters, C. J. (1997). Challenges in adaptive management of riparian and coastal ecosystems. *Conservation Ecology 1*(2), article 1. Retrieved from www.ecologyandsociety.org/vol1/iss2/art1.

Waltner-Toews, D., Neudoerffer, C., Joshi, D. and Tamang, M. S. (2005). Agro-urban ecosystem health assessment in Kathmandu, Nepal: epidemiology, systems, narratives. *EcoHealth 2*(2), 155–164.

Watzlawick, P., Weakland, J. H. and Fisch, R. (2011). *Change: Principles of Problem Formulation and Problem Resolution*. New York: W. W. Norton.

Wenger, E. (2004). Knowledge management as a doughnut: Shaping your knowledge strategy through communities of practice. *Ivey Business Journal 68*(3), 1–8.

Wheeler, A. (1973). From relativity to mutability. In J. Mehra (ed.), *A Physicist's Conception of Nature* (pp. 202–247). Boston, MA: D. Reidel Publishing Company.

Wittgenstein, L. (2010). *Philosophical Investigations* (P. M. S. Hacker and J. Schulte, Trans.). Oxford: Wiley-Blackwell.

Witzany, G. (2011). Can mathematics explain the evolution of human language? *Communicative and Integative Biology 4*(5), 516–520.

Wondolleck, J. M. and Yaffee, S. L. (2000). *Making Collaboration Work: Lessons from Innovation in Natural Resource Management*. Washington, DC: Island Press.

Index